Heredity and Evolution
in Human Populations

Second printing of
revised edition

"Every literate citizen, and not just every college student in search of a liberal education, should read these tantalizingly brief chapters. They will provide an indispensable orientation in the alarms of debate over the future of our genes. Geneticists, too, will find the volume not only a model of exposition worthy of emulation, but a source of many fruitful thoughts." — FROM A REVIEW OF THE FIRST EDITION IN *The Quarterly Review of Biology*

A century after Darwin's epoch-making work on evolution, Professor Dunn's book reminds us that "the meaning of evolution is change . . . evolution is not something accomplished, but something which is continually occurring. There is ample evidence that human evolution is occurring today." The author bears out this state-

Heredity and Evolution
in Human Populations

HARVARD BOOKS IN BIOLOGY

Number 1

Heredity and Evolution
in Human Populations

L. C. DUNN

Professor of Zoology Emeritus and
Former Director of the Institute for the Study of
Human Variation, Columbia University, New York

REVISED EDITION

HARVARD UNIVERSITY PRESS
CAMBRIDGE, MASSACHUSETTS

Copyright © 1959, 1965 by the President and Fellows of
Harvard College
All rights reserved
Revised Edition, 1967
Second Printing, 1968
Distributed in Great Britain by
Oxford University Press, London
Library of Congress card catalog number 65-11617
Printed in the United States of America

Preface to the Revised Edition

The first edition of this book, written in the summer of 1957, was addressed to people who wanted answers to questions about possible causes of changes in the biological constitution of human populations. It turned out that it was read by many who were looking not for easy answers or for final ones but for promising ones, that is, those that presented new views or opened new ways of study. This is the kind of interest that students have, and it was pleasant to find that the book proved useful to them.

When a new printing was called for it was the students' needs that I kept chiefly in mind, and the few changes deal with some new things which have been learned recently. I have not tried to be exhaustive in this respect, since anyone who grasps the rationale of this kind of study can go on and read for himself what has happened. Consequently I have added a selected bibliography of recent books which themselves contain lists of references to the scientific publications from which the illustrative examples in this book were drawn.

L. C. DUNN

Preface

Human heredity and human evolution present fascinating problems which touch every one of us. I think I am entitled to point this out after writing a little book about them because I was not originally of this opinion. Genetics, as I was introduced to it as a student, was an exact science because its statements could be verified experimentally; since human genetics could not be experimental, it must perforce consist only of applications of ideas derived from experiments with animals and plants. Ideas about evolution, on the other hand, were purely speculative and could not be tested at all by recourse to experiment. All this has changed, and I have changed my opinions too. Although I have spent most of my life in experimental genetics, I know that theories can be tested by other means than deliberately controlled experiment. Studies of human populations present many examples of this. One can imagine what arrangements of hereditary factors one ought to find in human populations and families if certain theories are true, and these arrangements can be observed. Theories of evolution can be tested both in this way, by observations of changes in populations in nature, and by deliberately designed experiments with rapidly reproducing animals, plants, bacteria, and viruses. The ideas derived from genetics and those from observations of species, races, and varieties of natural populations have now come to support and supplement each other. Inner consistency and relations between facts obtained from diverse sources, always a sign of maturation in a scientific field, begin to appear in the new field of population genetics. Its ideas are couched in general terms, applying to all populations with certain modes of mating and reproduction.

I have tried to select from these some ideas which seem to me basic and relevant, and to illustrate them by observations, most of them recent, made directly on human populations. I should not call the result a synthesis; that would be pretentious at this stage. It is rather an assessment of the promise inherent in these ways of looking at problems of human evolution. It is only problems of evolution on a small scale, modes of change in races and smaller communities to which I have addressed myself; microevolution is the term sometimes applied to problems on this level. Anyone wanting to read good discussions of the larger problems of evolution, should turn to G. G. Simpson's *Meaning of Evolution* and the more recent *Mankind Evolving* by T. Dobzhansky.

I have more debts to colleagues and friends than can be acknowledged individually. Individual chapters have been read by persons with knowledge and critical judgment of the matters in them, but not wanting to involve them in errors of commission or omission for which I am responsible, I must thank them anonymously. I should like, however, to record my gratitude to my colleagues in the Institute for the Study of Human Variation, especially Professor Theodosius Dobzhansky and Professor Howard Levene, for bearing with me in discussions of most of the problems treated in this book; and to my old teachers and friends under whose guidance I first became interested in heredity and evolution: Professor John H. Gerould at Dartmouth College and Professor William E. Castle at Harvard University. To my wife, Louise Porter Dunn, I am grateful for help and forbearance far beyond the call of duty.

NEW YORK, NEW YORK   L. C. D.
September 20, 1958.

Contents

Figures

Heredity and Evolution
in Human Populations

I

Variety

To a biologist, one of the most remarkable features of the human population of this planet is its enormous variety. Here among the 2500 million members of the one species, *Homo sapiens,* we find no duplicates, except those rare cases of identical twins which since they arose from one egg count as one individual. Literally each person is biologically unique and declares this fact not only in his obvious physical features, but in the individual properties of his blood and other body fluids, in the operation of his sense organs, and in numerous details of chemical constitution and behavior.

What is the biological meaning of this vast variety? The biological study of man and of other animal and plant species has established that it arises out of the interplay of two influences to which every living being is subject. One is the heredity transmitted to an individual by his parents through egg and sperm; the other is the varied conditions of life, the environments in which different individuals develop. The causes of hereditary variety are now known in outline. The transmission of heredity takes place by the passage of living units, genes, through the sex cells, egg and sperm, from parents to children. Reproduction consists essentially in the production by each gene of a replicate or copy of itself which passes into each new cell and into the new individual. The continuity of heredity and of life itself thus depends on the self-reproduction of the genes. Occasionally a copy is produced which is not quite identical, in its effects, with the parental form of the

gene. This kind of accident, of which the exact cause
is not known, is called a mutation. The mutated gene
then reproduces in its altered form so that two (or
more) forms of each such gene occur in the same
population. The thousands of such units which each
parent transmits to the children are then shuffled and
recombined in all possible combinations. The essential
features of the biological mechanism by which this occurs are now well understood.

The variety thus engendered is then acted upon by
the sifting effects of the varied environments in which
the species lives. Some combinations of traits prove to
be biologically more successful in certain environments
than in others. The end result is that local populations
show adaptations to the conditions under which they
have lived.

The two processes just outlined are those of *heredity*
and of *evolution*. In the last analysis they are both due
to one basic peculiarity of living substance, the ability
to reproduce itself by converting materials from the
environment (food) into its own specific configuration,
which is then perpetuated by the replication of hereditary units at each act of reproduction.

It is interesting to reflect that the chief clues to our
present understanding of each of these processes, of
heredity and of evolution, were obtained about 110 years
ago, and in relative independence of each other. On
the evening of July 1, 1858, the Linnean Society of
London met to hear two communications on the same
subject by two biologists, neither of whom was present.
The two papers bore the same title: "On the Tendency
of Species to form Varieties; and on the Perpetuation of
Varieties and Species by Natural Means of Selection."
The authors were Charles Darwin and Alfred Russell
Wallace. Each had hit upon the idea of natural selection
as the basic clue to the mechanism of evolution while
studying plants and animals in different parts of the

world, Wallace in Malaysia, Darwin in South America.

Darwin's book of 1859 elaborated the idea. Its full title was *On the Origin of Species by Means of Natural Selection, or the Preservation of Favoured Races in the Struggle for Life*. The chief facts on which the theory was based were, first, the enormous overproduction of offspring by all forms of life, of which only a few survived; and, second, the fact that not all offspring were alike but differed from each other by individual variations. The inference from these facts was that, since some of the variations were transmitted by heredity, they rendered their possessors and descendants better fitted to survive. These were the "favored races" which eventually give rise to new species. The historical changes through which living matter evolved and differentiated were seen as a result of forces inherent in nature itself; appeal to agencies outside of nature—supernatural forces—was thus made unnecessary.

It was a clear inference from the theory that if all life was part of the same historical process, guided by natural laws, then man was a part of it too. It was these last two features of Darwin's work that brought about the great intellectual revolution of the nineteenth century.

Even as Darwin's and Wallace's preliminary reports were being read, there was being initiated in a monastery garden in Moravia another train of events which was to lead to an understanding of the mechanism of heredity. There, in 1856, Gregor Mendel began his experimental search for the rules governing the inheritance of the characters by which varieties of the garden pea are differentiated. The report of his seven years of work was read to the Brünn Natural Science Society on the evenings of February 8 and March 8, 1865. No one before he wrote his report had conceived so clearly the essential problem of heredity in terms which could be tested by deliberately designed experiments. It was, in fact, Mendel's method of conceiving a theoretical model of a living process in precise terms, and then of testing it by simple direct experiments, which gave rise to a new era. Out of these beginnings came the science of genetics with its new insights into the nature of living matter, not least of which has been the elucidation of the mechanism by which evolution occurs.

HEREDITY

The essential feature of Mendel's discovery can be stated quite simply. It is that heredity is particulate, occurring by the transmission of discrete units known as *genes*. This was proved by observing the distribution among the descendants of single discrete traits in which the parents differed. Thus, in peas, a cross of a tall parent by a dwarf one both from true-breeding varieties produces only tall (hybrid) offspring. When bred together the hybrids produce three kinds of offspring in constant proportions: a quarter of them are tall and breed true like the tall grandparent; a half are tall but produce when interbred some dwarf offspring, like the hybrid tall plants; while a quarter are dwarf and breed true like the dwarf grandparent. Mendel explained this by supposing that pure tall plants transmit through each reproductive cell (*gamete*) a unit, or gene, leading to the tall habit of growth; the dwarf plants transmit similarly a gene for dwarf habit in each gamete. Union of these two gametes produces a hybrid with one of each kind of gene, although the tall gene has the prevailing influence on growth (known as *dominance,* but not a constant or essential feature of heredity). The crux of Mendel's theory lay in his explanation of what happens when such a hybrid forms its reproductive cells. Then, said Mendel, the two genes in the hybrid act as sharp alternatives, one kind (tall) entering half the gametes, the other kind (dwarf) going to the other half. Every gamete thus receives one and one only of each pair of alternative genes. When hybrids are crossed together the gametes of one parent (½ tall, ½ dwarf) and the gametes of the other (½ tall, ½ dwarf) meet in random fertilization. These random combinations would be ½ tall × ½ tall = ¼ tall-tall; ½ tall × ½ dwarf + ½ dwarf × ½ tall = ½ tall-dwarf (hybrids); and ½ dwarf × ½ dwarf = ¼ dwarf-dwarf. This constant distribution can be obtained only by assuming discrete units which separate sharply (segregate) when the gametes are formed. When present together in the same individual they do not blend or contaminate each other. Each gene retains its own integral unitary character. This rule was derived from each of the seven pairs of alternative traits in peas studied by Mendel and is the general rule in all organisms studied, from viruses and bacteria to man.

We know now that the hereditary material transmitted by animals and plants generally consists of thousands of discrete units, genes, each of which follows the behavior described above, although we can observe this fact only when a gene has assumed by mutation two alternative forms, such as those leading to tallness or dwarfness. The genes are structures distributed in a definite order in larger bodies of the nucleus of each cell, the *chromosomes,* which occur in a number that is characteristic of each species. The pea plant, for example, has regularly 7 pairs of chromosomes; in man there are 23 pairs (Fig. 1). The chromosomes are visible under the microscope; individual genes are too small to be seen by present methods. This is not surprising, considering that chromosomes, each of which may contain hundreds or thousands of genes, are so small that a billion or more of them may fill no greater bulk than that of a drop of water.

Research in the first 20 years following the rediscovery of Mendel's laws in 1900 established the general

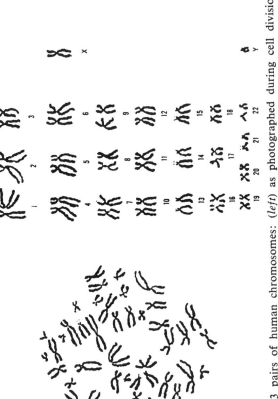

Fig. 1. The 23 pairs of human chromosomes: (*left*) as photographed during cell division (metaphase); (*right*) the members of the pairs have been cut out and arranged in order of size. The cell is from a normal man and so contains 22 pairs of autosomes and an unequal pair consisting of an X and a Y chromosome. (From O. J. Miller, B. B. Mukherjee, and W. R. Breg, *Trans. N. Y. Acad. Sci.,* 24:372, 1962.)

architecture of the genetic system in the chromosomes. Genes located in different pairs of chromosomes were found to recombine at random when the reproductive cells are formed, entering freely into all possible combinations in accordance with Mendel's second principle of independent assortment. Genes located in the same pair of chromosomes also enter into all possible combinations, but this process may be retarded, in terms of generations, by the tendency of genes in the same chromosome to retain their associations with each other. The strength of this tendency, known as linkage, is a measure of their nearness to each other in the chromosome, and this has permitted the construction of "gene maps" for many rapidly reproducing species of plants and animals.

More recent research on the experimental analysis of heredity has shown that these rules are of general application to the behavior in reproduction and hereditary transmission of all organisms from viruses and bacteria to man. The rules derive ultimately from the particulate molecular structure of the elements of heredity—the genes—and from the manner in which these reproduce themselves and get transmitted to progeny. Genes can be conceived as units or segments of the long chainlike molecules of deoxyribosenucleic acid (DNA) which compose the chromosomes. These molecules consist of successions of four different kinds of parts (nucleotide bases) which can occur in an enormous number of different patterns or orders of arrangement, each related to an effect on a primary bodily process such as the synthesis of a specific protein, an enzyme, which governs the development or the functioning of a part or process of the body. Such a pattern constitutes a genetic code or set of instructions for the construction of a new individual. It is copied when the DNA molecules replicate themselves during the formation of the reproductive cells. It is thus that descendants get copies of the parental kinds of DNA molecules. An essential feature of heredity is thus the making and transference of duplicate molecules.

The self-copying process, which occurs thousands of times in the growth of an individual and in the production of its thousands or millions of germ cells, is subject to accident. Rarely, the copy of a gene is not quite like its original and subsequently replicates in its new form, making possible new effects on the traits of descendants and new combinations of genes. Such events, known as *mutations* are the ultimate source of the hereditary variety characteristic of all species. The effect of mutation which produces a variety of gene forms is enormously enhanced, in bisexual organisms like man, by another essential feature of heredity. The genes from the two parents enter into new combinations each time an egg or sperm is formed and each offspring is thus likely to have a unique combination of genes. It is this variety due to mutation and recombination upon which natural selection and other evolutionary forces act in forming varieties, races, species, and other natural categories.

It is now known that the probability of error in the copying process, that is, the frequency with which mutations take place, is influenced by forces impinging upon the living cell from its environment. Ionizing radiations such as x-rays and emanations from radioactive elements, heat, light of different wavelengths, and a variety of chemical compounds are known to affect the mutation rate. The basic step in the origination of variety is thus probably subject to physical and chemical laws, although the detailed application of these to the mutation process is not yet known.

What all organisms inherit, therefore, is a pattern of self-reproducing units, each with its specific properties. The fact that like usually begets like is due to the stability of these units; the fact that hereditary unlikenesses arise shows that the stability is not absolute. It is the latter property which gives the opportunity for change and introduces the variety which permits organisms to become adapted to particular environments.

Another general principle derived from the study of heredity has often been misunderstood. It is this. What we inherit is genes, true enough. But this does not mean that we, or any of the higher animals, inherit our bodily or mental traits or characters, as such. These arise during the course of a long process of development, in which not only the genes in the fertilized egg (from which development originates) but the conditions of the environment play important roles. The genes determine the forms which this response to the environment takes. They set the norms of reaction, the ranges of responses which are possible. Thus it is not true to say that we inherit a disease such as diabetes. The disease results from the interaction between a particular genetic constitution and an environment which lacks a particular substance, insulin. But when that substance is supplied, that particular genetic constitution does not inevitably produce the disease. The essential difference between the person who needs insulin and the one who does not lies in the different reaction norms which they have inherited. A similar relation, to a greater or lesser degree, exists between what is inherited and what is expressed in the traits (collectively known as the phenotype) of the individual.

What we inherit is the result of the self-reproduction of genes determining modes of reaction to the environment and of the transmission of copies of parental genes in all possible combinations to the offspring.

EVOLUTION

The essential meaning of evolution is change. As applied to living beings it means the process by which changes occur in the biological constitution of a population. In modern terms, biological constitution may be defined as the assemblage of hereditary elements, genes, characteristic of the individuals of the population. A *population,* in the biological sense in which we shall use the word, is a group of individuals who have received their genes from a common pool established by reproductive communication, mating, among antecedent members of the population. The simplest human model is the population of a small isolated island where marriages occur only between residents and everyone is related by common descent. Within such a closed population the variety of genes present when the population began would maintain itself in the descendants unless augmented by mutation or changed by emigration or immigration, or by differential survival of the possessors of different gene assortments, or by other influences to be discussed later. Any of the latter agencies might with time alter the biological constitution of the population. We should say then that *evolutionary changes* had occurred in the population and refer to the influences that had caused them as *evolutionary forces.*

The most extensive unit of population is the *species.* Members of a species are interfertile, and by intermating the members have access to a common pool of genes. The genes of each species tend to be confined within it by the absence or rarity of matings between members of different species. Thus different species are separated from each other by gaps; they are populations which are reproductively isolated. The human species is separated by this kind of barrier from all other primates.

Within the species *Homo sapiens* all members are potentially capable of mating and producing offspring with all other members. This fact gives all men access to the community of genes which constitutes the patrimony of the species.

In practice, of course, genes could not circulate freely through a species as widespread geographically as ours and consisting of communities fully or partially isolated from each other by natural or cultural barriers. Each of us could hardly have the equal probability of marriage with any other member which might obtain in a small island population. It is more likely that we shall marry someone near at hand, someone speaking the same language, professing the same religion, with respect for the traditions in which we have grown up; in short, propinquity and similarity in social and cultural background will largely determine our marriage patterns.

Thus it will come about that any species population will usually consist of smaller population units within which matings are most likely to occur. These may come to contain samples of the common gene pool of the species which differ from one another in the frequency with which some of the genes are represented. This is usually the first step in the biological differentiation within the species, the first step in evolutionary change. In man, it results in the formation of races, which took their origins from geographic separations among segments of the species population.

Evolutionary changes may in the beginning be of relatively small dimensions, consisting in the divergence of populations only in the relative frequencies of some of their genes and characters. Within some West African tribes, for example, 25 percent of the members of populations inhabiting a river bottom have a gene for sickle-cell anemia, while the same gene is found in only a few percent of members of mating units of the same tribe living at higher altitudes. Inhabitants of Sardinian villages only a few miles apart show wide and significant differences in the proportions of their inhabitants who have one particular blood-group gene.

It is at this incipient stage that evolutionary changes and the forces that produce them can be most readily identified and studied in living human populations. That out of steps like these arose the historical changes which constitute evolution in the larger sense is highly probable. If the smaller problems, often referred to as microevolution, seem limited and modest as compared to the grand strategy of organic evolution to which Darwin addressed himself, we may at least have the satisfaction of studying history as it happens today, and by identifying the forces which mold human populations and small communities we may feel firm ground under our feet in contemplating the longer vista of the evolution of man.

II

The Principles of Heredity
Applied to Populations

The question now to be addressed is how the hereditary
constitution maintains itself in a population of animals
which reproduce by outbreeding. Man is such an animal.
In large communities we usually marry someone from
a family unrelated to our own; and even in small communities, the continued mating of close relatives, that
is, inbreeding, is subject to restriction by law or social
custom, so that marriage partners are usually unrelated
in the sense that an ancestor common to both man and
wife is usually too remote to be recorded or remem-
bered. This means that the hereditary constitution of
each individual, consisting of a large number of different genes, will be derived from a large number of
different ancestors. If there are no ancestors common
to the mother and the father, for, say, ten generations,
then any present member of such a population had
$2^{10} = 1024$ *different* ancestors ten generations ago.

Before we can appreciate what this kind of outbreeding implies for the maintenance of variety in human
populations, we shall have to understand something
about the actual mechanism of heredity.

The first proof of the existence of elementary units of
heredity was given by Mendel in 1865. His experimental
evidence and the theory derived from it are given in
detail in his paper (published 1866) and this is as clear
an exposition of the essential idea as anything published
since. Instead of repeating here the details (which can

be consulted in Mendel's paper or in any elementary book on genetics), I shall use as a simple illustration of the main principle the manner in which a single human gene is transmitted. We learn about this by tracing the descent of some simple characteristic in which people differ. One such sharp and clear difference is revealed when we put a sample of red blood cells, each from a different person, in certain specific testing fluids known as *antiserums*. We may, for example, take cells from a person arbitrarily designated as person M, inject them into a rabbit, and later draw from the rabbit some blood serum which, after appropriate treatment, we now find will cause the red cells of person M, when placed in it, to form clumps, a reaction known as *agglutination*. Cells of another person, whom we shall call person N, do not agglutinate when placed in this serum but remain free as they normally are. We say that the rabbit has formed *antibodies* which have a specific effect on M cells but not on N cells. Now with cells from person N we produce, in a rabbit, antiserum with antibodies specific for N.

If now we take some red blood cells from each of a number of different persons (chosen at random in an American town, for example) and add some cells from each individual sample to a small amount of each of these two different testing fluids (antiserums), which we shall call arbitrarily anti-M serum and anti-N serum, respectively, the following results are observed. Cells from certain individuals form clumps (agglutinate) in anti-M serum but fail to do so in anti-N serum. Cells from certain other individuals clump in anti-N serum but fail to do so in anti-M. Cells from all the remaining individuals form clumps in both anti-M and anti-N serum. This shows that the population consists of three kinds of people with respect to this reaction. The first kind may be described as M people, the second as N, and the third as MN.

We now apply this same test to parents and their children, and find the results shown in Table I. It is

Table 1. Results of matings between parents classified for their MN blood groups.

Mating type	Parent 1 (father or mother)	Parent 2 (mother or father)	Classification of children		
			M	MN	N
1	M	M	All	None	None
2	N	N	None	None	All
3	M	N	None	All	None
4	MN	N	None	½	½
5	MN	M	½	½	None
6	MN	MN	¼	½	¼

obvious that the MN blood types of the children are determined by those of the parents, that is, by heredity.

In interpreting the mode of inheritance we may assume that all gametes of M people (egg or sperm) transmit something that stands for the property M. Similarly, N people transmit a property N in every gamete. The union of an M gamete with an N gamete (mating type 3) produces an individual with both properties. Now it is the distribution of the two properties among the offspring of such MN people which gives the clue to the theory. They clearly transmit M to half their children and N to the other half (mating types 4 and 5) and to none do they transmit both. We may then assume that MN people form gametes of only two types in equal number, ½ with M and ½ with N. If this is so, then the unions of these two types that should occur in matings of MN by MN are shown in Table 2. The children from this union should appear in the ratio ¼ M, ½ MN, ¼ N. This is, in fact, just about the ratio found when children from large numbers of matings of type 6 are tested. (There will, of course, be statistical

Table 2. The principle of segregation illustrated by matings of two heterozygotes MN MN.

Father's sperm	Mother's eggs	Children will be
M	M	M
M	N	MN
N	M	MN
N	N	N

fluctuations in different samples of people, but the averages tend to be 25 percent M, 50 percent MN, and 25 percent N.) We may take this as confirmation of the basic assumption that something determining this blood type occurs in two alternative forms, and that each form is found in half the eggs or sperm of an MN individual.

Now we must attach some names to these assumed elements and the individuals who contain them. The element responsible for this blood type we shall call a *gene*. The two alternative forms of it we shall call *alleles*. One allele is to be called M and the other *m* (not-M). (It is always best to designate forms of the same gene by modifications of the same letter, M and m, or M^1 and M^2, in order that members of pairs may be recognized as such at once and not confused with members of other pairs of alternatives.) The individual that has like alleles is a *homozygote* (a *zygote* is the cell or individual arising from the fusion of two gametes, egg and sperm); the individual with unlike alleles is a *heterozygote*. The genic description of an individual is its *genotype*. The genotype of an M individual is *MM*, of an N individual *mm*, of an MN individual *Mm*. The character expressed by the individual, what it looks like (or tests like in this case), is called its *phenotype*. In the case of the MN and most of the other blood types, each genotype has its own unique phenotype, but this is not true in many other cases. There is in man, for example, a gene, one form of which is responsible for albinism, the virtual absence of dark pigment. Here there are three genotypes but only two phenotypes, as follows:

genotypes:	*A A*	*A a*	*aa*
phenotypes:	normal	normal	albino

Since the allele *a* has no visible effect in the heterozygote, it is referred to as a *recessive* allele, a term introduced by Mendel. Its alternative *A* is referred to as the *dominant* allele. Alleles with dominant effects are designated by capital letters, recessive by small letters. In many cases, as in MN, there is no dominance, both alleles expressing themselves in the phenotype of the heterozygote.

Now we come again to the essential feature of the theory: the sharp separation of the alleles always into *different* gametes. This basic postulate of genetics is usually referred to as Mendel's principle of segregation.

The opportunity for identifying a gene by its behavior at segregation is given only when it exists in at least two distinguishable forms such as *A* and *a*. The existence of genes is thus an inference from the ways in which alternatives (alleles) with different effects are distributed among the offspring.

We must, of course, ask how it comes about that some genes exist in two different forms. We may assume that at one time there was but one form, say *M,* and that at some later time a unit reproducing as *M* gave rise by an accident—a miscopying of a parent gene by a daughter gene—to a new form *m* which thereafter remained stable, reproducing as *m*. This event is known as *gene mutation*. Events of this sort have occurred under observation, and in fact means are at hand, such as application of radiations, heat, and various chemicals, for making mutations occur much more frequently. But just when the allelic forms of most human genes arose is usually unknown.

WHAT AND WHERE ARE THE GENES?

Now just what are these elements which we have inferred from the alternative behavior in inheritance of traits like the M blood-group reaction or albinism? The evidence clearly indicates that they are actual physical elements of submicroscopic size, borne in bodies which can be seen under the microscope. In each one of our cells there are 23 pairs of such bodies, known as *chromosomes* (literally colored bodies) since the DNA of which they are chiefly composed takes up certain colored dyes which make them visible under the microscope. Each species has a characteristic and constant number of chromosomes in each cell: Mendel's peas had 7 pairs, the corn plant has 10 pairs, the house mice have 20 pairs, and we (*Homo sapiens*) have 23 pairs.

Each individual receives at conception 23 of his set of 46 chromosomes from his mother and 23 from his father. Twenty-two of these form equivalent pairs in each cell of the body: two of chromosome pair 1, two of pair 2, and so on. The twenty-third pair consists of unequal members and is related to the sex of the individual. Females have two of this kind of chromosome, known as the sex or X-chromosome, but only one or the other of them is active in any single body cell. Males have an X and a Y chromosome, the latter carrying genes determining maleness.

In what follows we shall treat all 23 pairs as equivalent in transmission and take up later (in Chapter IV) the relations of the X-Y pair to sex.

When an individual forms reproductive cells, eggs or sperm (collectively known as sex cells or *gametes*), one member of each of these pairs of chromosomes, and only one, enters each gamete. Since each gamete contains only half the number of chromosomes characteristic of the body cells of the species, the process by which the gametes are produced is known as the *reduction division,* a part of the general process of *meiosis,* the division of cells preparatory to the production of eggs in the female, of sperm in the male (Fig. 2). It is the reduction division which distinguishes this process from the reproduction of other cells by which growth occurs (known as *mitosis*), in which each chromosome produces a replica of itself. This is the essential act of all reproduction and by it the full number of chromosomes is maintained in all cells other than those which form the gametes. Since chromosomes arise only by the replication of ancestral chromosomes, this process provides, as we have already seen, the essential mechanism of continuity of living substance.

At the reduction division it is purely a matter of chance which member of any chromosome pair is distributed to any single cell, that is, whether it is the member derived from the individual's father or from his mother. To make this clearer, consider only two out of the 23 pairs; two of the chromosomes would be of paternal derivation, 1^P and 2^P, and two would be maternal, 1^M and 2^M. We know that these will appear in the gametes of that person in these combinations: 1^P2^P, 1^P2^M, 1^M2^P, 1^M2^M, and that all these combinations are equal in number. This must mean that the chance that any pair member, say 1^P, will enter any gamete is $\frac{1}{2}$, since half the gametes get this member; the other half get 1^M, so that the chance of 1^M is $\frac{1}{2}$. Similar chances obtain for each member of any other pair. It is also a rule of the game that the distribution of the members of one pair is independent of distribution of the members of any other pair; the pairs exert no influence on each other. Hence the chance for the combination 1^P2^P is $\frac{1}{2} \times \frac{1}{2}$ or $\frac{1}{4}$; for 1^M2^M it is also $\frac{1}{4}$; that is to say, the chance that any gamete will transmit two chromosomes of paternal origin is $\frac{1}{4}$; similarly for two of maternal origin; while, since there are two ways of getting a gamete with one paternal and one maternal

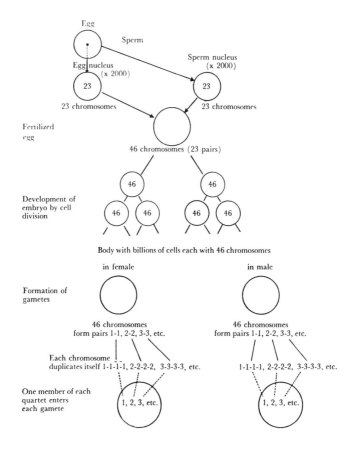

Fig. 2. The chromosomes in human reproduction and development.

chromosome, half of the gametes will have this mixed constitution (Fig. 3). The same game can be played by flipping two coins which fall head or tail by chance. The chances for all four combinations—head-head, head-tail, tail-head, and tail-tail—are equal, namely, ¼ two heads, ½ one head and one tail, and ¼ two tails.

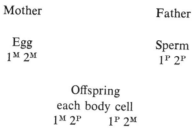

Mother   Father

Egg   Sperm
$1^M 2^M$ $1^P 2^P$

Offspring
each body cell
$1^M 2^P$ $1^P 2^M$

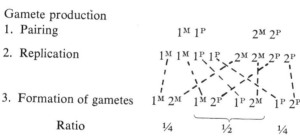

Gamete production
1. Pairing $1^M 1^P$ $2^M 2^P$

2. Replication $1^M 1^M 1^P 1^P$ $2^M 2^M 2^P 2^P$

3. Formation of gametes $1^M 2^M$ $1^M 2^P$ $1^P 2^M$ $1^P 2^P$

   Ratio   ¼   ½   ¼

Fig. 3. The behavior of the chromosomes when the gametes (egg and sperm cells) are formed, taking pairs 1 and 2 as representative of all 23 pairs.

Now the rules of this game are obviously the same as the rules governing the distribution of the alleles of one gene. Two alleles, like or unlike, are present in each individual, but only one in each gamete. If an individual has unlike alleles, say M and m (that is, is heterozygous), half the gametes get M and half m.

Our first guess might be, therefore, that the same event that decides the separation of the alternative members of a pair of chromosomes into two different gametes also decides the separation of the pair of alternative alleles. And this guess proves to be correct. The mechanism of the reduction division (that is, of meiosis) provides the physical mechanism for the segregation of alleles.

It would take us too far from our present purpose to cite in detail the proofs that the chromosomes actually consist each of a long succession of different genes, hundreds or thousands of them in linear order.

Alleles of the same gene occupy identical positions (loci) on one chromosome pair, one allele on one member, the other allele on the other, opposite to the first when the chromosomes align themselves in pairs as they do at the reduction division. Genes on different chromosome pairs are distributed to the progeny at random with respect to each other. Thus if $A-a$ is a pair of alleles on chromosome pair 1, and $B-b$ a pair on chromosome pair 2, then the double heterozygote *AaBb* forms the four types of gametes in equal number: ¼ *AB*, ¼ *Ab*, ¼ *aB*, and ¼ *ab*. In matings of two such heterozygotes, both sets of like alleles *AABB* and *aabb* appear each in about ¹⁄₁₆ of the offspring (¼ *AB* × ¼ *AB*, and so on). If the parents of the heterozygote had been *AAbb* × *aaBB*, then the gametes *AB* and *ab* would represent recombinations of these alleles which had been in different parents. (For example, if *A* stands for woolly hair, as found in some European families, *a* for straight nonwoolly hair, *B* for patches of white on forehead and chest found in other European families, and *b* for normal pigmentation, we should find after intermarriage between members of these two kinds of families some offspring inheriting both woolly hair and spotted skin *AB*, a recombination of these alleles.) The rule for independent assortment (Mendel's second and less general principle) applies when alleles belonging to different pairs appear in the gametes of a heterozygote in all possible combinations with equal frequency. Different genes on the *same* pair of chromosomes do not recombine at random but tend to retain in the offspring the associations they had in the parental chromosomes. This tendency is known as linkage and quantitative estimates of the degree or strength of linkage permit estimates of the relative distances between gene loci on the same chromosome and consequently "gene maps" of chromosomes can be constructed. This has not proceeded very far with human chromosomes and in any case is beyond the scope of this book.

With this background of insight into the transmission mechanism of genes and chromosomes we return to our first problem. What is the effect on human populations of generally inheriting chromosomes and genes from many different ancestors? Suppose we apply the rules we have found to one of those 1024 ancestors in the tenth ancestral generation of our hypothetical "outbred" person and ask what is the chance that one of his two chromosomes of pair 1 will be found in any one descendant today. If the chance of transmission of this chromosome is ½ for each act of gamete production and 10 such acts are involved, then the answer is $(½)^{10}$ or $\frac{1}{1024}$; while the chance that one of the descendant's first chromosomes has come from some *other* ancestor is $\frac{1023}{1024}$. There is no chance at all that both of the descendant's first chromosomes could have come from the same ancestor since all matings are between unrelated people, who thus could not have got another first chromosome from this ancestor. We may ask, however, what chance there is that both a first and a second chromosome of the descendant have come from the same ancestor. If it is $\frac{1}{1024}$ for the first and $\frac{1}{1024}$ for the second chromosome, the chance of getting both together is $\frac{1}{1024} \times \frac{1}{1024} = \left(\frac{1}{1024}\right)^2$. For one of *each* of the 23 chromosomes of that ancestor to appear together in one descendant it is

$\left(\frac{1}{1024}\right)^{23}$, which is so small a number that we say the chance is negligible that just one of each of his chromosomes will appear in any one descendant. Since the combination of individual peculiarities of that ancestor depends upon particular properties of his chromosomes, the genes, it appears that under this system of mating his particular combination of genes would be dissipated, virtually never to reappear in his descendants.

We have played this game of chance only to show that in human reproduction and in the mating system generally (but never exactly) followed in man (there are never the infinite number of unrelated potential mates that are assumed in the universe in which pure chance operates) there exists a mechanism for the generation of variety in the sources from which the hereditary material of each individual is derived. This may seem to prove too much, for our original question concerned the maintenance of the hereditary constitution, and we have found one method by which it may be dissipated.

There is no paradox here, however, for two qualities are maintained by this system of chromosome distribution. First, each individual receives from his ancestors a full and exact complement of just 23 pairs of chromosomes—human chromosomes—within which must be embedded all that makes him human, for these chromosomes are all that he receives biologically from his ancestors. The bodies of his ancestors have disappeared; but before they did, the one thing that accounts for the continuity of their life and qualities in all their descendants occurred: their chromosomes and genes replicated themselves. This self-replication of genes is continuity itself. Heredity, variation, evolution—all these are consequences of the duplication of the genes at each cell division, at the formation of each gamete. Rather than bone of the ancestor's bone or blood of his blood, the descendant has the chromosomes of his chromosomes in living continuity. The effect is clear to see: in the descendants appear all the essential zoological features of the ancestors. The system also keeps open the door to novelty and variety. If errors in the replication process —mutations—have occurred, the probability is great that these will be distributed in different combinations among the different ancestors. It is this which accounts for the fact that, within the sameness produced by self-replication, there is never complete identity in the genetic patrimony which different descendants receive.

One should think of a typical pair of human chromosomes as having the constitutions shown in Fig. 4. Here the gene symbols are arbitrarily chosen. One may suppose them to represent effects of diverse kinds on pigmentation, blood antigens, hair form, stature, and so forth. The upper chromosome of the pair will have come to this individual from one parent, the other from the other. Most of the alleles in one parent are shown as different from those in the other, expressing the fact that the two chromosomes will have been derived from different ancestors, hence will probably be unlike. Most humans are known to be heterozygous for many of their genes.

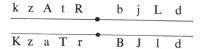

Fig. 4. Hypothetical arrangement of genes in a single pair of human chromosomes.

The effect of recombination either between genes on different chromosomes or between those on the same chromosome is to increase enormously the initial variety due to gene mutation by permitting the alleles to enter into all possible combinations. Take the case of the two pairs of alleles cited above. A mutation in one kindred

gave rise to two kinds of persons in respect to hair form
—woolly or straight. A mutation in another kindred
gave rise to persons with normal or with spotted skin.
By intermarriage between these kindreds and gene recombination, opportunity is given for the appearance of
four kinds of persons: (1) with woolly hair and normal
skin, (2) with straight hair and spotted skin, (3) with
straight hair and normal skin, (4) with woolly hair and
spotted skin. With three pairs of different alleles, *Aa,
Bb,* and *Cc,* there would arise eight visibly different
combinations of characters in which one allele shows
dominance (as the above do) over the other. The
numbers of combinations rise exponentially with the
number of gene differences. If n is the number of genes
differentiated into two alleles each, the number of different possible phenotypes is 2^n where one allele is dominant, 3^n where both alleles express themselves. Ten gene
differences thus give rise to $2^{10} = 1024$ phenotypes
(with dominance), or to $3^{10} = 59,049$ (without dominance); and ten, in terms of actual experience with
genes which occur in two allelic forms, is a very small
number. This enormous variety will be generated in any
outbreeding population in which mutations have occurred regardless whether some of the changed genes
are linked or not. Linkage will slow down, in initial
stages, the attainment of the maximum variety but in
time it should express itself fully wherever recombination occurs at all.

GENES IN POPULATIONS

The heaviest going in respect to outlining principles
and theory is now over. How can they help us to understand the chief problems we set out to study: the maintenance of biological variety in human populations and
the identification of forces that cause change in the
variety in an evolutionary sense?

The foundations of our present views about these

questions were laid not long after Mendel's work was rediscovered and widely confirmed early in this century. An English statistician jumped to the conclusion that dominant genes ought to spread in populations, recessive ones to diminish in commonness. An English mathematician, Hardy, quickly showed (in 1908) that his statistical colleague was wrong. Hardy proved, using only Mendel's original postulate of segregation and some simple algebra, that in large populations in which matings occur at random with respect to genotype (as they do in man) any two different alleles such as A and a which are equal in their effects on probability of survival will retain their relative proportions indefinitely so long as certain specified conditions obtain. The same proof was provided independently in the same year by the German physician Weinberg, whence the idea is referred to as the Hardy-Weinberg equilibrium principle.

The proof is as follows. Assume that two alleles A and a of one gene occur in one population. Some members will then be AA, some Aa, and some aa. What will be the proportions of these genotypes in the population and in its descendants? The chief factor determining this will be the relative number of alleles A and a in the population. Harking back to Mendel's original proof we know that when both parents are Aa we can predict the genotypes of the offspring by applying his first principle, namely, each parent Aa forms equal numbers of gametes A and a (and only these). Random union of these at fertilization produces ¼ AA, ½ Aa, ¼ aa, as shown in Fig. 5a.

The proportions of A and a alleles are equal in the population of parents, ½ A and ½ a; they are also equal in the children, ½ A and ½ a. The gene frequency has not changed. The population of children, if allowed to mate at random in all possible mate pairs, will again produce the same proportions of genotypes

among their children since the gene proportions are the
same. The important idea of random mating allows us
to treat all representatives of this gene pair Aa in such
a population as members of a common pool of genes
to which every person (by reason of equal probability
of mating with any other person of opposite sex) has
equal access.

$$
\begin{array}{c}
\text{Eggs} \\
\text{Sperm} \left\{
\begin{array}{c|cc}
 & \frac{1}{2}\ A & \frac{1}{2}\ a \\ \hline
\frac{1}{2}\ \overline{A} & \frac{1}{4}\ AA & \frac{1}{4}\ Aa \\
\frac{1}{2}\ a & \frac{1}{4}\ Aa & \frac{1}{4}\ aa
\end{array}
\right. \quad (a)
\end{array}
$$

$$
\begin{array}{c}
\text{Eggs} \\
\text{Sperm} \left\{
\begin{array}{c|cc}
 & q\ A & (1-q)\ a \\ \hline
q\ \overline{A} & q^2\ AA & q(1-q)\ Aa \\
(1-q)\ a & q(1-q)\ Aa & (1-q)^2\ aa
\end{array}
\right. \quad (b)
\end{array}
$$

Fig. 5. Showing the method of computing the outcome of
random combination of genes in a population: (a) when
two genes A and a are present in equal numbers; (b) when
they are present in the ratio $qA : (1-q)a$.

Now obviously, of any two alleles such as A and a,
one may be very common, the other (like albinism)
very rare. The allele proportions, considering a variety
of different genes, may take any values whatever. We
should therefore let "any proportion" be represented by
a symbol, and say that the proportion of A alleles in a
population is q; the rest of the alleles, the a ones, will
be equal to $1 - q$. We can now get a general solution
to the question: what is the proportion of genotypes

AA, Aa, and *aa* in a population in which *a* has the frequency 1 — *q?* We do just as we did in getting the numerical answer, namely, reckon out the consequences of random fertilization of the two kinds of eggs and sperm, as shown in Fig. 5*b*.

The answer is that under random mating the proportion of genotypes in the offspring will be $q^2AA + 2q(1 - q)Aa + (1 - q)^2aa = 1$, which is the form of the equilibrium formula or equation for general use, the Hardy-Weinberg equation.

Those interested in carrying the question one generation further can easily prove that the proportions of alleles in the above generation are $qA:(1 - q)a$, that is, the same as in the previous generation. The population generated by random union among these will have the same proportions as the first generation and so will all subsequent generations. That is why the population is said to be in equilibrium: it retains its allelic variety in constant proportions. The rule applies to each pair of alleles separately, each having its own equilibrium value, and to all combinations of different genes. If a second independent pair of alleles was present in proportions *r* and 1 — *r*, the gamete proportions would be given by the product

$$[qA + (1 - q)a] [rB + (1 - r)b]$$

and the genotype proportions in the population by the random combination among these gametes, as above.

So here we have the basis for biological stasis or constancy of hereditary variety in a random breeding population. But before applying these ideas too widely, and especially before examining the question of how the qualities of such a population could change to a new equilibrium condition, we should look more closely at some of the assumptions on which the formulation rests, for it is these that hold the clues to both constancy and change.

THE CONDITIONS OF CONSTANCY

The basic assumption is, of course, Mendelian segregation, the existence of units, which remain intact in all combinations and of which the alternate forms, such as *A* and *a*, appear in equal numbers among the gametes from the heterozygote *Aa*. This is a postulate subject to strict test by observation and experiment. Proof of its general validity has been given repeatedly, in man as well as in animals and plants generally. If it had not been so proved we should still have to deduce it from the fact that in many cases three phenotypes within different populations (such as M, MN, and N blood types) do actually occur in about the proportion q^2MM: $2q(1 - q)Mm$:$(1 - q)^2mm$, even when M and m have very different frequencies. Occasional departures from the 50:50 segregation ratio have been noted. If these are not due (as they often are) to random statistical fluctuations which will even out in time, but to actual excess of one allele over the other such as *a* over *A* among eggs or sperm from *Aa*, then the allele appearing in excess will increase in the population and an evolutionary change will occur. Since no such cases are known in human genetics, and the few instances in other animals are clearly exceptional and not well understood, we may safely use Mendelian segregation as a cornerstone in our thinking about equilibrium conditions in populations.

A second assumption is that the proportion of alleles is not affected by mutation that is predominantly in one direction, say from *A* to *a*, or vice versa. If there was a greater probability of change from *A* to *a* than the reverse, a pressure of mutation favoring *a* would be built up and this allele would tend to increase. We know that genes occasionally change by mutation; their stability is relative rather than absolute; but in general the mutation rate of most human genes is low enough to justify the use of the assumption.

One assumption is certainly not generally valid. No population is of unlimited size; and yet this is the only condition under which events can happen purely at random. But, as we get reasonably close approaches to the outcome expected by pure chance in tossing coins or throwing dice *if* we do it a large number of times, just so our assumption that matings and gamete combinations occur at random holds reasonably well in populations which do not fall below some lower limit of size. What the actual lower limit will be will depend on how we intend to apply the theory and what conclusions we want to draw from events that appear not to fit the theory. Let us say therefore that the theory applies to large populations, without specifying how large.

In small closed populations of a few hundred people, matings between relatives, cousins for example, will occur by chance more frequently than in large open populations. This would tend to increase the proportions of rare homozygotes *AA* and *aa* and decrease the heterozygotes *Aa,* since genes coming from a common ancestor are more likely to be alike. But cousin matings or other forms of inbreeding will not, of themselves, change the *proportions* of the alleles.

However, such changes might occur purely by chance if, in a small population, the few possessors of some gene failed to find mates or to leave offspring. Suppose there are only a few genes for red hair in a small isolated population. The risk of losing that gene, quite aside from the desirability or undesirability of having red hair, is obviously greater than in a large population with the same proportion of red hair genes. The greater risk of extinction of one allele with consequent fixation of its alternative, due to accidents of sampling in small populations, is clearly a cause of change which has to be considered. It is referred to as *random drift*.

In addition to the large size of population assumed to permit the operation of chance, we assume also that randomness in mating is not interfered with by mating preferences or patterns which lead toward assortative mating. There is some tendency for spouses to resemble each other in certain respects (in height, for example) more than would be expected on a chance basis; but unless differences in height reflect the action of the same genes which determine, say, a blood-type difference, mating might be at random with respect to the second character even though it were not with respect to the first. To persons descended from the romantic traditions of western Europe, the term random mating itself seems to engender some revulsions; but, in fact, that is our system with respect to most human genes. How could it be otherwise when we have so little knowledge of what those 23 chromosomes of our prospective life partner contain?

Another most important assumption is that the alleles in the equilibrium formula should be equally subject to natural selection. Whenever *A* and *a* have unequal chances of being passed on to subsequent generations, this will not be so and consequently there will be changes in the gene proportions.

The essential meaning of natural selection is just that one biologically distinguishable kind of plant or animal or man should contribute more genes than another to succeeding generations. Suppose, for example, that *AA* and *Aa* have normal phenotypes while *aa* has a biological handicap, such as an anemia, which causes all *aa* children to die before reaching reproductive age. Each such death carries two *a* genes out of the population and, other things being equal, this gene will eventually be reduced to a proportion determined by the mutation rate of *A* to *a*. This form of anemia will thereafter appear only as a result of mutation, a rare event. Or *aa* or *bb* or *cc* people may be favored by higher fertility, by being preferred as marriage partners and getting married earlier or having more children, in which case the allele favored by such selective processes will increase in proportion to its alternative. Merely adding years to the end of the life span of persons with a particular gene combination would not increase the frequency of such genes, since years after the end of the reproductive period would not result in any change in the proportion of the alleles passed to the next generation, and this is all that matters.

Finally, the Hardy-Weinberg formulation has to assume that, for maintenance of constancy in any one population, the population will not lose or receive migrants who have genes in proportions which differ from those in the population. If gentlemen on another planet preferred the blondes on ours to the brunettes on theirs, and could get ours away, the gene frequency of blondness on earth would decline. Or if a large number of universal blood donors, those of blood group O, should be called up in an emergency and fail to return to the population, the supply of the gene O would be depleted. It is of course only migration which is biologically selective that matters.

This last condition about migration reminds us that the ideal population in which constancy or equilibrium of gene frequencies is maintained has to be reproductively isolated from other populations with different gene frequencies. The history of human populations warns us that this condition must frequently have been violated, evolutionary changes in one population having been caused by primary changes like mutation which had happened first in other populations.

What we have then is a theoretical model or framework of the expected working out of certain assumptions. Its real existence (to make a justifiable paradox) is only in the unreal or abstract world of probability. Its greatest usefulness is to allow us to compare events in our own real world of flesh and blood with those in the ideal or abstract model. In this way we have a chance of detecting and measuring the degree of those departures due to biological events such as mutation, natural selection, or limitation in size of actual populations which result in change in the equilibrium of the population, that is to say, in evolution.

III

Methods of Evolutionary Change in Human Populations

The opportunity for change, in an evolutionary sense, in a human population is given whenever there is a failure of one or more of the conditions on which constancy of its gene proportions depend. Opportunity by itself does not mean change; the tendency in populations is to continue in the state of balance or equilibrium which represents its adjustment to past conditions. In order for the population to move to a new state or level of equilibrium the existing state must be upset for a sufficient number of generations that a new balance leading to a new adjustment may be attained.

The social and political changes in human history have accustomed us to thinking of societies as resultants of balances of forces. Conservatives and radicals moderate each other's effects, whether it be political issues at stake, or the substitution of a new kind of bow and arrow for an old, or of atomic power for more traditional sources of energy. Eventually the balance is upset; societies and their technologies change.

Most comparisons between social and biological evolution cannot safely be carried beyond this general occurrence of equilibrium. The forces involved are different. We say that history does not repeat itself, meaning usually that the conjunctions of men and the conditions they face at different times are likely to be unique. But in the slow process of human biological evolution unique events will play a lesser role because it is man's biologically conservative nature which must respond; and many generations of repeated exposure to the changed evolutionary forces are required for that. We have only to remember those 23 pairs of chromosomes with their freight of good genes that enabled him to cope with past environments to appreciate the risk involved in jettisoning something old for some new and unique biological will o' the wisp.

Let us list again the evolutionary forces which play on man's biology and examine each to see why repetition and steady pressure are essential conditions to give them effect.

Mutation. Mutation supplies the primary source of variety on which all the other forces operate. Yet a single new mutation in a population will not thereby change it to a new level of equilibrium. The usual fate of a single new mutation will be to be lost. Suppose, in a population of several thousand individuals, a single gene changes from condition *A,* representing, let us say, the capacity of the red blood cells to hold 100 molecules of oxygen (O_2) in combination under certain conditions, to condition *a,* representing the ability to hold 105 molecules. Condition *a* is more efficient than *A,* but will only be exhibited when two mutated genes are present in the same condition, *aa;* that is, *a* is recessive. As a single event, *a* will appear first in a single individual *Aa,* with the prevailing rate of oxygen transport, that is, having no advantage over *AA.* The only mates for *Aa* are *AA* individuals. If no mate is found, then *a* disappears from the population. Even if a mate is found, there is always a 50-percent chance that *a* will be lost. The opportunity has been presented but not realized. If *a* has some beneficial effect in the *Aa* individual, its chance of persisting is increased, but because it is unique at origin it still runs the risk of extinction from purely random causes. In general, the chance for the new condition to establish itself will depend upon the repetition of the opportunity. Biologists refer to this as *mutation pressure,* due to recurrent mutation.

Random drift. Random fixation or loss of an allele due to drift in small populations could happen only after the previous occurrence of gene mutation, and is thus a dependent or secondary possibility. But even where mutation has occurred, the mere passage of a population through a bottleneck of small size only gives the opportunity for random changes in gene proportions to occur. To be an effective means of evolutionary change, even in this limited sense, repeated opportunities of this sort would have to occur. Each such opportunity might be purely temporary, unless the population in which it had occurred happened to encounter conditions favorable to its new state. The opportunity presented by mutation and random fluctuation in the frequency of the gene could be seized only where it coincided with an environmental opportunity as well.

Gene flow or migration. Similar comments might be made about changes in the genetic constitution of a population due to migration or flow of genes into or out of a population by hybridization or intermarriage. A single red-haired girl dropping into a large population of black-haired people would to be sure cause some excitement, but as an unrepeated event would hardly ripple the gene pool. A single intermarriage of, say, a European into an African or an American Indian tribe would have little chance of effecting a permanent change in the character of a whole population. Such events would have to be repetitive to give them effectiveness as agencies in the long historical process by which populations change.

The cards would thus appear to be stacked against evolutionary effects occurring by such agencies and especially against changes due to single or unique events.

Natural selection. When we come to selection, the outook at first glance would seem to be brighter. Why should not a character which confers on its possessors greater fitness to cope with the environmental conditions and which repeatedly occurs in a large population quickly establish itself and displace an alternative and less fit character? Here we have to hark back to our basic view of heredity as outlined in Chapter I and recall that what is inherited is a way of responding to the environment. We should expect to find in a population those ways of responding which have proved most successful in the past. Any change in this pattern is likely to upset these responses unless some change in external conditions accompanies the change in heredity. These two are not likely to coincide by chance. Using the metaphor of lock and key by which the norm of reaction was pictured, if locks and keys both vary at random with respect to each other, then it is unlikely that a better fit between them will happen by chance. This does not mean that it cannot happen, only that it will be rare for characters conferring greater fitness to appear. Advantage or fitness is not an attribute of the hereditary elements themselves but results from the relation between the genes and the environment in which the population lives. Examples of this from human populations will soon be given in detail, but the idea is so important that I cite now a dramatic instance of its truth from recent work with populations of bacteria.

The bacterial population is composed of individuals each consisting of a single cell. These usually reproduce simply by dividing into two (fission) without the sexual process which in higher organisms involves the union of two different cells, one from each parent. The descendants of a single bacterial cell each inherit the whole genotype of the parent cell, hence, unless mutations occur, are all alike, forming a *clone,* the asexual or vegetative products of one individual. It has been shown, nevertheless, that mutation is continually occurring at a low rate in such populations, and that a single population may consist of two or more genotypes as in man and other higher organisms. Moreover, in some bacterial populations processes comparable to mating and exchange of genes have been discovered, and this has made possible the proof that the genetic system is composed of genes which change by random mutation and that these obey the same rules of Mendelian inheritance which were first discovered in higher plants and animals.

Now bacterial populations are remarkable in the speed with which they can reproduce. The divisions by which a single cell becomes 2, 4, 8, . . . , 2^n may succeed each other at intervals of 20 minutes, so the population may expand with great rapidity. If there are, for example, two genotypes in a bacterial population, differing in the efficiency with which they can extract from their environment the energy needed for growth, the results of competition between these should be quickly evident if one counts, at the beginning of an experiment and after 50 or more generations, the relative numbers of cells of the two different types.

In this way it has been shown that, from populations of bacteria most of which are killed when an antibiotic such as streptomycin is added to the culture medium, there may be obtained, relatively quickly, strains which are highly resistant to streptomycin. The facts now established are that mutations conferring resistance to streptomycin occur repeatedly in strains which have not been exposed to the drug. The rate of mutation is low, just a few occurrences per million cells. In the absence of streptomycin these may have no effect on the population, and such individual mutations would not persist. But put such a population in an environment containing streptomycin and a remarkable change in its descendants occurs. The cells with the usual or normal gene (for susceptibility) are killed or prevented from reproducing; those with the mutant gene are favored, and even though initially so rare as to have a frequency which one might consider negligible, their descendants quickly overgrow the culture and the "normal" type disappears except for rare mutation in the opposite direction, namely from resistance to susceptibility to the drug. The population has, as we say, adapted itself to the new environment.

Shall we conclude in this case that the mutation to resistance was "beneficial"? The answer is clearly no if we judge it before the application of streptomycin, when the resistant cells had no advantage, clearly yes if we judge it in the new environment. That fitness or advantage derives from a specific relation between the genotype and the environment is shown by the fact that a strain which has become resistant to one antibiotic may still be susceptible to other bactericidal agents such as ultraviolet light or various chemicals. Strains resistant to these latter agents may be obtained by growing them in contact with ultraviolet or the appropriate chemical.

This may be taken as a model of the operation of natural selection. Out of a variety of genotypes generated by mutation and recombination those are selected (that is, pass their genes to later generations) which are fitted to cope with the environment.

Evolutionary changes of the type described above have been observed to occur not only in microorganisms but in animals and plants which reproduce with sufficient rapidity to permit an experimenter to follow them for a number of generations. The outcome is the appearance of a new state of equilibrium in the population by which it has become adapted, fitted, to its environment. One may assume that, in general, populations have evolved their adaptedness to their environments by similar processes of natural selection, but it is seldom possible to specify the particular genes which have changed their frequencies in response to particular environmental factors. However, some examples have emerged from recent work with human populations which tend to validate the application to man of evolutionary principles derived from studies of other forms. I propose to describe some of these observations as illustrative of the ways of studying microevolution in man and as presenting problems which have in themselves a good deal of human interest.

HUMAN VARIETY IN DIFFERENT ENVIRONMENTS

As we survey the kinds of human beings indigenous to different parts of the world we cannot fail to be struck by certain obvious relations between the biological characteristics of a population and the physical conditions under which it lives. It seems somehow appropriate that the natives of tropical Africa should have black skins and ability to withstand heat and humidity, and that in the subarctic regions we should find squat, compact people with flat faces, able to live successfully in extreme cold. And yet when we try to explain how this adaptive relation arose between the population and its environment we are hard put to it to specify the actual elementary steps connecting a biological trait such as black skin with an environmental agent such as high temperature or humidity. Biologists of an earlier generation entertained an explanation that now seems rather vague. It was that black skin was originally an acquired character that became inherited. It was assumed that the skin color was the direct adaptive response of ancestors to the effect of hot sun; the altered character was then transmitted by inheritance to the descendants.

The inheritance of acquired characters no longer satisfies biologists familiar with the facts of heredity, mainly because there is experimental proof of the origin of changed characters by mutation and of Mendelian inheritance which renders such a vague and unproved hypothesis unnecessary. Modern research, on the contrary, seeks to identify the elementary steps in the process of adaptation of a population to its environment, difficult as this usually is with human beings.

Thus it is evident that skin color is governed by genes which determine the norm of reaction to the environment, genes which have diverged by mutation and are transmitted according to the Mendelian rules. Segregation can be observed in the children of marriages between racial hybrids, as between Europeans and Negroes; several genes are involved but the results conform to the general Mendelian hypothesis. The problem thus becomes to explain how a number of genes making for dark skin color attained high frequency in African and other dark-skinned populations, and lesser frequencies in other populations.

The same problem is posed by biological differences between populations generally. What led to the concentration in Asiatic populations of genes for the kind of eye fold, of hair form, of the form of face and color of skin characteristic of the Mongoloid peoples? That there are different genes influencing these structural peculiarities is known from the study of racial hybrids. What was it in the environment in which the characters of such a population took their present form that enabled them to survive and to supplant alternative characters? What were the selective agencies in their environments that played the role of streptomycin in the bacterial populations cited above?

Asking the question in this specific form reveals at once how difficult it is with our present information to answer it for the gross morphological features by which we customarily apprehend local differences between populations. Seldom is such a difference as that between black and white skin observed to arise in a single step. True, albinism occurs as a rare recessive gene in most populations, but it has never given rise to an albino population, because albinism is accompanied by disadvantages to its possessor in any human environment in which man is likely to live—poor eyesight, sensitiveness to sunburn, and other disabilities. Smaller departures from the usual skin color found in a population are not so penalized; in fact, blondes are said to be more successful than others in certain preliminaries to mating. So the problem probably involves not one or a few genes with large individual effects but many, each with less marked effects. And these are difficult to identify because their effects are not separated by steps sharp enough to judge by eye and are, moreover, blurred by environmental influences such as the tanning effects of sun and wind and the varying states of health. Moreover, the outwardly obvious effects of the gene or group of genes may not be the most important ones which influence the population's chances of survival in a given environment. People with black skins have also more pigment in the iris, which screens the retina from strong light; their vision is better at very high intensities of illumination. The dark melanin pigment in the deep layers of the skin and in the eyes is produced by the action of enzymes (ferments or catalysts) on colorless compounds. Greater efficiency of this pigment-forming process may mean also greater efficiency of connected chemical processes in the body under certain environmental conditions.

The chains of cause and effect connecting the variations in man's body which characterize large populations —size, skin color, hair form, skull shape, and the like— with particular environmental factors must often be long and complex and devious. This is what might be expected in a complex mechanism like the human body, in which the external features are the outcome of delicately adjusted and interdependent physiological processes. We shall have to know more about these basic processes before we can spell out their relations to the genes at one end of the chain and to the external environment at the other. We can be sure that such relations exist, for adjustment to an environment is a prime necessity for every living thing. The search for such causal relations in the case of racial variations in man will be aided by the kind of hopeful speculations recently indulged in by three physical anthropologists, Coon, Garn, and Birdsell. In a provocative little book (*Races,* 1949) they have suggested some reasons why particular physiques may be more successful in certain environments, for example, why a compact body form, with lower heat loss, is better fitted for life in a cold climate than a longer, more slender one. Most of these connections remain to be tested, and, like the work of physical anthropologists generally, they deal with phenotypes, with external characters of a complex nature rather than with the genes, the distribution of which must hold the clue to the riddle of man's adaptive responses to the varied environments.

It has recently become possible to use new methods, based on directly identifiable human genes, to study the problem of evolutionary adaptation in man. Suppose we ask, not why a particular physique, but why a particular gene is common in some parts of the world but not in others. Here for the moment we put aside such questions as what the population looks like and ask the simpler question: how many genes of a particular sort can we count within it; and compare populations first only in this way. This is the kind of simplification which Mendel introduced when he shut his eyes for a moment to the appearance of his plants as a whole and concentrated first on one trait at a time. The reason behind our attempt with man is just the same as his: to apprehend processes, rules to which particles or elements, not wholes, conform. Eventually, of course, sense must be made of what we find in terms of wholes, since it is the fate of living men in particular environments which is our problem. In the end what we find from the study of gene distribution must square with what is discovered by other ways of studying man and, as will be shown later, there are already signs of agreement between the conclusions of traditional physical anthropology and of those who use the new methods of gene frequency.

The study of genes in human populations has received in recent years a series of stimuli which have acted literally like shots in the arm. The first came with Landsteiner's discovery in 1900 of the four "classic" blood groups, A, B, AB, and O, which became available for use in population study when they were later shown to be determined by alleles of one gene which had different frequencies in different populations. A strong booster shot was the discovery of the Rh blood-type genes by Landsteiner and Wiener in 1940 and the proof by Levine of the relation between these genes and the death of newborn babies from hemolytic jaundice. These discoveries were rapidly exploited in America and Europe and especially in Great Britain where the widespread national blood transfusion service, developed during World War II, led to the identification of blood-type genes in hundreds of thousands of persons and to the discovery of several additional genes of this sort. The blood types of millions of persons in all parts of the world are now known for a dozen or more genes.

Then in 1949 came the discovery by the American chemist Pauling and his colleagues that a gene previously identified as the cause of a fatal anemia—sickle-cell anemia—actually altered the form of the hemoglobin molecule on which the oxygen transport mechanism of the blood depends. This brought to bear on the study of this gene, and now on many others like it, the great arsenal of method and theory which had been developed by physical chemistry; and knowledge of the distribution and effects of many genes for a variety of abnormal hemoglobins expanded almost explosively. Today the condition of these genes is known in several hundreds of thousands of persons in populations in different parts of the world.

In the 1940's, the Italian investigators Silvestroni and Bianco tested in Italy alone over 100,000 people for the presence of another gene affecting the red blood cells, that for thalassemia or Mediterranean anemia, which in homozygous form produces the fatal disease known as Cooley's anemia. Investigators in the United States, Greece, Turkey, India, and the Far East quickly extended our knowledge of this gene and its distribution in the world's populations.

In addition to the genes known to affect the structure of hemoglobin and of the red blood cells, others have now been identified by their effects on components of the blood serum. One of these controls the structure of the haptoglobin proteins. Populations in many parts of the world have now been shown to be polymorphic with respect to the alleles of this gene, which have different frequencies in different populations. It appears now that other genes affecting serum proteins (transferrins, gamma globulins) are also maintained in polymorphic state in many human populations.

A similar expansion in knowledge is now occurring with respect to an X-chromosome gene which affects the ability of the red blood cells to carry out certain important chemical transformations. This gene is referred to as G6PD, from the initials of the enzyme (glucose-6-phosphate dehydrogenase) which it affects. It is widely distributed in populations of African and Mediterranean descent.

The extensive studies of the geographical distributions of these genes affecting the character of the blood led to descriptions of human populations which resemble in one important respect those built up by ethnographers from observations of other biological characters. This is that populations vary in the relative frequency with which particular traits are found in them. Just as African populations south of the Sahara are marked by high frequencies of very dark skins, so they are marked also by the world's highest frequencies of one of the Rh genes (R^0) and by high but variable frequencies of people with the sickle-cell gene.

Blood-group gene *B* is commoner in Asia than elsewhere but absent in the indigenous populations of the Americas. The blood-type gene *M,* used as an example in Chapter II, is common in some populations, rare in others; similarly, the genes for thalassemia, sickle-cell anemia, and many others show wide differences in frequency in different populations. The world distribution maps of these genes will be discussed in more detail later. At present I want to point out only the nature of the problem presented to those interested in human evolution. How was this state of affairs reached, in which the world's populations appear when looked at by the gene frequency method as a kind of patchwork mosaic of differing degrees of commonness or rarity of the same genes? How have the elementary forces of mutation, selection, random drift, and migration interacted to produce such a mosaic? How can these factors be studied, evaluated, and made to throw some light on the causes of evolutionary change in human populations?

Although the most extensive new information pertains to the distribution of the blood-group genes in races, tribes, and communities, it is the picture now emerging from the rapidly growing knowledge of the hereditary anemias which has given the most encouragement to students of evolution. For, in the case of sickle-cell anemia and of thalassemia, the distribution of the genes appears to be related to tangible factors of the environments of the populations in which they occur. The evidence is compelling that natural selection is the paramount force determining the frequencies of these genes.

SICKLE-CELL ANEMIA

The red blood cells are normally circular, discoidal, in shape; but in 1910 an American physician, Herrick, found red cells of bizarre forms, some of them sickle-shaped, in the blood of a West Indian Negro suffering from anemia. The condition turned out to be inherited. The primary clue to the mode of inheritance was the discovery that both of the parents of persons suffering from sickle-cell anemia (in which most of the red cells are deformed) always had what has come to be called "sickle-cell trait," that is, some of the red cells of the parents became sickle-shaped when deprived of oxygen. When the children (sibships) born to such parents were examined they were found to consist of three categories (phenotypes) in the following proportions: ¼ non-sicklers (normal), ½ with sickle-cell trait, and ¼ with sickle-cell anemia. The latter usually die from the disease in early life. Matings between persons with sickle-cell trait and persons with normal red cells produce children half of whom have normal red cells and half have sickle-cell trait. These are just the results to be expected if the difference between the abilities to produce sickle cells or normal ones is determined by a single pair of alleles of one gene—*Si,* sickling, and *si,* normal. Persons with sickle-cell trait are evidently heterozygotes *Si/si,* normal people are *si/si* homozygotes, those with sickle-cell anemia are *Si/Si* homozygotes. The three phenotypes produced by matings between persons with sickle-cell trait evidently correspond to the three genotypes in the basic Mendelian proportions given above: ¼ *si/si,* ½ *Si/si,* ¼ *Si/Si.*

For many years these genotypes were detected by examining a few drops of blood under appropriate conditions and observing the absence or presence of sickle cells. But in 1949 it was shown by Pauling and others that hemoglobin, the protein which gives the color to the red cells and which can be extracted from them in solution, differed in these three genotypes. When a solution containing different proteins is placed between two poles of an electric field, molecules move toward one of the poles at speeds determined by the electric charge on the molecule. Since different molecules, in this case different proteins, differ specifically in their electric charges, this method (known as electrophoresis) may be used to separate and identify by their different speeds of movement two or more different classes of molecules in the same solution.

When hemoglobin from a person with sickle-cell trait was tested by electrophoresis, it was found to consist of two different kinds of molecules. About half of them (roughly 60 percent) moved with the speed characteristic of the single kind of hemoglobin found in normal nonsicklers; the other fraction moved more slowly and became separated from the normal hemoglobin. The hemoglobin from persons with sickle-cell anemia (homozygous for the abnormal gene) all moved with the abnormally low speed of the slow molecules found in sickle-cell trait. To complete the proof, hemoglobin from normal people was mixed in equal parts with hemoglobin from sickle-cell anemics. Under electrophoresis the mixed solution separated into two parts, one characteristic of the normal, the other of the abnormal or slow kind.

Thus it was shown for the first time that a gene could determine which specific kind of molecule was to be produced by the body in one of the primary activities of living matter, the synthesis of proteins. The paper announcing this discovery was appropriately entitled "Sickle-cell anemia, a molecular disease." Pauling was later awarded a Nobel prize for this and other discoveries.

This pioneer work had immediate technical consequences, since it now became possible to test for molec-

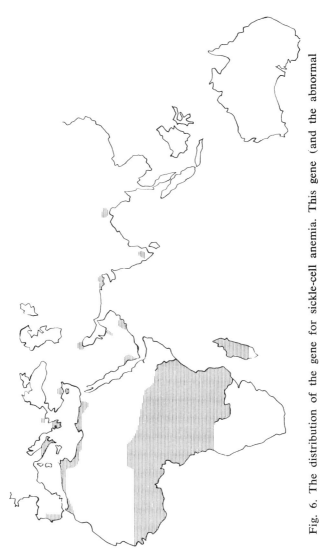

Fig. 6. The distribution of the gene for sickle-cell anemia. This gene (and the abnormal hemoglobin which it produces) has been found in all the populations around the Mediterranean (except in Spain), in Africa south of the Sahara and north of the Zambesi River, in the Middle East, in South India, and in Assam.

ular variety not only the hemoglobins but other proteins of the blood (in the blood serum, for example) and other tissues. A great variety of these, associated with genotypic differences, has been revealed in populations in different parts of the world. The theoretical consequences will eventually be more interesting, since now a pathway has been opened, in man himself, for exploring the mechanisms by which the machinery on which life depends (the proteins) is put together. Already it has been shown that the change from the normal gene to its sickle-cell allele, which arose first by mutation, does in fact change one electric charge at one point in the molecule, just as predicted by the theory derived from electrophoretic mobility. The essential feature of the mutation responsible for sickle-cell anemia (hemoglobin S) has been shown by Ingram to be due to the substitution of a single one out of nearly 150 amino acids in each of the two chains in the half molecule of normal adult hemoglobin. Another mutant gene, that for the abnormal hemoglobin C, found chiefly in West African populations, was similarly shown to be due to substitution of a third type of amino acid at the same point in the molecule as that affected by the sickle cell gene. The same kind of combined chemical and genetical study has shown this kind of localized unitary change to be associated with the origin of other gene-controlled abnormal hemoglobins and has made it possible to construct a simple theory explaining the evolution of the genes controlling the structure and functioning of this most important human protein. These changes in gene function are in turn probably traceable to localized changes in the pattern of structure, the genetic code, in the molecule of deoxyribose nucleic acid which constitutes the material basis of hereditary transmission.

Now, with the nature of the gene difference between sicklers and nonsicklers revealed at least in outline, let us see how the presence of one or the other or both of these two alleles of one gene is related to the lives of the populations in which they occur. The sickle-cell gene was first discovered in a Negro, and central Africa south of the Sahara and north of the Zambesi river turns out to be the world focus of high frequencies of this gene, although, as the map (Fig. 6) shows, it does reach appreciable frequencies around the Mediterranean, especially in Sicily, Greece, and Turkey, in southern Arabia and in India, and of course in persons of African descent now living in the New World. One of the peculiarities of its distribution in Africa is the wide fluctuations in its frequency which occur even within members of the same racial group. The highest frequencies (up to 45 percent of persons with sickle-cell trait) are found in lowlands, especially along rivers and at their mouths, and in general in those regions in which malicious malaria is prevalent, that is, where nearly everyone gets infected by the parasite (*Plasmodium falciparum*) which is transmitted from person to person by anopheline mosquitoes, and which is responsible for this form of malaria. In other parts of the world, too, populations with many sickle-cell genes inhabit malarial areas, as around Lake Copais in Greece, known since the time of Strabo the Greek geographer (b. 63 B.C.) as a center of the disease and now a focus of the highest frequencies of sickle-cell trait in Europe.

It would be a seeming paradox to assume a causal connection between malaria and sickle-cell disease, for this would mean that a disease which by itself causes high mortality (malaria) leads to an increase in the proportion of genes responsible for a molecular disease which is fatal to homozygotes. Before we face this paradox let us see what should normally happen in a population in which many individuals are heterozygous for this gene. Matings between such heterozygotes would be frequent. If a quarter of the adult population were heterozygotes, then a sixteenth of all matings occurring

at random ($\frac{1}{4} \times \frac{1}{4}$) would be between two heterozygotes. From each such mating, a quarter of the chil-
dren would be homozygous for the sickle gene and most
of these are known to die before having children to
whom they would pass on the gene. Each such death
carries two sickle genes out of the population, and the
frequency of these genes should decline toward the van-
ishing point unless their supply is augmented by some
other factor such as mutation.

We can reach the same conclusion by applying the
methods used in deriving the Hardy-Weinberg formula
(Chapter II). If a quarter of the adults have each one
sickle gene and one nonsickle gene of this pair (being
Si/si) while three quarters have two nonsickle genes
(si/si) then the proportion of these genes in the population (the gene pool shared by random mating) should be
$\frac{1}{8}$ Si : $\frac{7}{8}$ si, that is, the frequency, q, of the sickle genes
should be $\frac{1}{8}$ or 12.5 percent. The offspring produced
by random matings in such a population should then be
as follows:

$$q^2 = (0.125)^2 = 0.015625 \text{ } Si/Si \text{ sickle}$$
$$\text{homozygotes}$$
$$(1—q)^2 = (0.875)^2 = 0.765625 \text{ } si/si \text{ nonsickle}$$
$$\text{homozygotes}$$
$$2q(1—q) = 2(0.125 \times 0.875) = 0.218750 \text{ } Si/si$$
$$\text{Total } \overline{1.000000} \text{ heterozygotes}$$

It has been found that the proportion of sickle homozygotes among infants shortly after birth does in fact
conform to the proportion expected from the Hardy-
Weinberg formula.

But what if all the sickle homozygotes die from
anemia before the reproductive age is reached? Then in
the adult population above (after depletion by deaths)
the sickle genes comprise only 11.1 percent of all instead of 12.5 as in the parents. This result is reached in

the following way. The frequencies of the genes in the survivors are: *Si*, 0.21875; *si*, 0.21875 in the heterozygotes; *si*, 1.53125 (2 × 0.765625) in the normal people; the sum of these frequencies is 1.96855, of which 0.21875 is 11.1 percent. The sickle-gene frequency has fallen in one generation to 89 percent of its former value, and this decline would be expected to continue, slowing down as the gene approached extinction. This is, in fact, the usual fate in populations of genes which kill their possessors when homozygous. Such genes are known as lethals, and under natural selection they are expected (and known in some experimental populations of animals) to be reduced to low frequencies which can be maintained only by new occurrences of mutation of a normal gene to its lethal allele.

And so we are faced in the case of the sickle gene (a nearly complete lethal) with a further paradox: it should not exist at all in the frequencies with which it is found in many populations! It should not exist, that is to say, unless some other factor supplies in each generation new sickle-cell genes to replace those lost by death, or protects from extinction those already present.

Authorities are agreed now that it is not mutation which is the principal source of the high frequencies of sickle genes. The rate of mutation would have to be improbably high (about 1 per 1000 genes compared to about 1 per 100,000 for other human genes); and the mutations would have to be limited just to certain populations in which the high frequencies occur. Moreover, there is no direct evidence of an unusual mutation rate for this gene.

So we had better turn to the first paradox, the possible protection or encouragement of sickle genes by malaria. For this hypothesis there is some direct evidence. Allison in 1954 found, in a West African population with a high proportion of sicklers, fewer young children infected with malaria among those with sickle-cell trait (28 percent infected) than among those without it (46 percent infected). He then infected with malaria 30 adult volunteers from one West African tribe. Of these 15 were sicklers and only 2 got malaria, while of the 15 nonsicklers all but 1 got malaria. Other investigators in Africa and the United States have made similar tests and obtained a variety of results, not all indicating the increased resistance of sicklers to malaria. The question is evidently complicated by the existence of acquired immunity conferred by previous early exposure to malaria, but one fact appears to be established: children in malarial areas who have one sickle gene have fewer malarial parasites than children without such a gene. It is thus highly likely that mortality from malaria will be lower in persons heterozygous for the sickle gene. In malarial areas it seems literally to pay off in terms of survival to have a sickle gene. Heterozygotes may have an advantage estimated by Allison at about 25 percent over normal homozygotes. This would explain the higher proportion of sicklers in adults than in infants of the same population: the nonsicklers have suffered higher mortality.

Assuming this to be so, we can now make some sense out of the distribution of this gene which always attains its highest frequencies in areas in which for long periods malaria has been endemic. We can also predict what should happen to sickle genes in such areas. Instead of being eliminated as in nonmalarial areas by the death of homozygotes from sickle-cell anemia, they should now be retained in an equilibrium in which loss from such deaths is balanced by increase due to the larger number of children produced by the protected heterozygotes. There is thus no need to appeal to high mutation rates to replenish the supply of sickle genes.

Although all the facts about this interesting relationship are not yet available and final interpretations cannot be reached, it stands as a conceptual model for the action of natural selection in a human population. Wherever selection favors a hybrid or heterozygote, both of the alleles, in this case the one for abnormal hemoglobin and its nonsickling allele, may be retained even though both types of homozygotes are disadvantageous. This produces the type of population structure referred to as balanced polymorphism, which appears to be an evolutionary adaptive device in populations of animals, plants, and men.

I have refrained from referring to the nonsickle allele as normal, for it too is deleterious in malarial areas. "Normal" is clearly a relative term, always to be judged in the context of the conditions obtaining in the population, and, since we seldom or never know these with any completeness, it is usually not a useful term for students of evolution, even though it may be for those concerned with health and disease.

We are led by such a model also to see some evolutionary reason in the existence of balanced polymorphism. In spite of the price which the population may have to pay in the earlier death of most homozygotes, a population in which the three genotypes Si/Si (sickle anemics), Si/si (sickle trait), si/si (non-sickle) have reached equilibrium proportions will show in this respect an adaptation to its environment and will have its best fitness in an evolutionary sense. Moreover, the equilibrium should be able to shift and adapt itself to maintain the fitness of the population as the environment changes.

Thus, we can conceive that in populations long exposed to malaria, the occurrence of the sickle gene may have been one of the factors that enabled them to cope with such an environment. Having tided them over such a period, what may now be expected to happen to this gene if malaria is eradicated, as it has been recently in Europe and America? If the relation assumed above is correct, we can confidently predict that the incidence of the sickle gene will decrease over a number of generations until it reaches the rare state in which the only new cases of sickle-cell anemia will trace to mutations of recent occurrence. This should happen unless heterozygotes are favored by some other factors, in addition to resistance to malaria, of which at present we are not aware.

And now, since man's inventions have joined the list of evolutionary forces through the use of DDT to eliminate malarial mosquitoes, what should happen if a cure for sickle-cell anemia is found and the homozygotes which now die should be enabled to contribute their quota of children and genes to succeeding generations? Obviously the fall in the frequency of this gene would be arrested in proportion to the relative number of sickle genes thereby enabled to be passed on, and we should have to be prepared to continue to cure the disease as a permanent obligation of the social system which proposes to maintain itself in health. Or we could dissuade or prevent all persons known to harbor the gene from having children. We could order them sterilized, a measure which has been sparingly employed against other ills, especially mental ones, but which encounters difficulties in societies in which individual freedom is prized. Or we could teach the use of contraceptive techniques and urge all heterozygotes to employ them.

Let us be still more fanciful (as evolutionists who must take long views have a good right to be) and imagine that a new radioactive element, one brought into existence, say, by a new kind of atomic bomb, increases the probability of the event by which a gene is caused to mutate to a lethal allele, such as the sickle allele. We might not know of the existence of such a connection, but in our eagerness to make the world safer by exploding bombs in it, we might load the atmosphere with this element. This too would become an evolutionary force, and a new balance between mutation and selection would have to be attained, that is, some future population would have to adjust to an environment in which the frequency of mutation had been increased.

I have dwelt at some length on this one gene not only because we have some good facts and theories about it but chiefly because we meet in this case most of the questions which have to be considered when we think about evolutionary processes in human populations. The chief factors involved in the adaptation of a population are always its genetic constitution(and this means the proportions within it of genes of different kinds) and the conditions to which adjustment has to be made, in short, our old friends heredity and environment. Evolution can, in fact, be defined as change in gene proportions in response to challenges from the environment. It must involve changes which have the kind of durability associated with continuous elements, genes; otherwise the adjustment is transitory and not part of a continuous historical process. This means that it is usually *past* environments which have determined the constellation of genes and the arrangements of these in present-day populations. To account for the present distribution of frequencies of sickle genes we have to assume that mutation must have been producing such genes in the past at a rate sufficient to give the chance for one such mutation to "catch on" when the opportunity arose for it to serve some function connected with adaptation in the population. We have to imagine what the elements were which combined to give the opportunity. In this case these include geological processes which produced the physical environments conducive to success of the anopheline moquitoes; the origin of obligate parasitic relations between the malaria plasmodium organism on the one hand and its two hosts—man and mosquito—on the other; historical processes, perhaps basically economic in origin, which drove man to colonize such regions; to say nothing of the causes which produced the social organizations which held men in gene-sharing, mating, communities. In evolution it is always populations which adjust, groups within which patterns of genes take form which enable the group and eventually the species to take advantage of the variety of environmental opportunities which present themselves.

THALASSEMIA

A gene resembling that for sickle cells in its distribution and dynamics within populations is that for thalassemia, an hereditary anemia first recognized in populations from the shores of the Mediterranean Sea (thalassa, Greek for sea) but now known to be distributed in a wide band from Spain and North Africa to Indonesia as well as in some Negroes. The genetic facts are simple. In children with the fatal disease known as Cooley's anemia, or thalassemia major, the red blood cells are small and abnormal and most of the hemoglobin is of the type found normally only in the embryo (fetal hemoglobin) and in infants, being replaced in early life by adult hemoglobin. These cells are inadequate to carry out the normal oxygen transport of the blood and are subject to rapid destruction in the spleen, which enlarges to cope with the extra load. Such children usually die before the age of puberty. Parents of such children have themselves a characteristic blood picture of mild departure from normal red cell forms, diagnosed as microcytemia or thalassemia minor. These persons with thalassemia minor are heterozygous for a gene *Th,* while patients with thalassemia major are homozygous for this gene *Th/Th.* From matings between heterozygotes three classes of children are born : ¼ nonthalassemic, ½ with thalassemia minor, ¼ with thalassemia major who die. The gene *Th* may be regarded as a complete lethal and should therefore be subject to reduction in the population to the low frequencies due to new mutations to this gene.

But in many populations the thalassemia gene is far from rare. The most extensive population surveys have been carried out in the towns and villages of the Po delta in northeastern Italy. Here in the neighborhood of Ferrara the proportions of heterozygotes in some villages reach 20 percent. If 20 percent of all adults have one such gene, the total proportion of such genes in the marrying population must be about 10 percent, and 1 percent $(.10^2)$ of all children should therefore die from this anemia. Physicians in this region have been aware for generations of the fate that awaits many of the children, but have been powerless to avert it. The deaths were known to be concentrated in certain families. We know now that this was because, in such families, marriages had taken place between persons each carrying one thalassemia allele, that is, with thalassemia minor. In Italy the disease and the gene which causes it have a peculiar geographic distribution. Peak frequencies occur in the Po delta, at the "tip of the boot" in Calabria, in Sicily, and in the coastal areas of Sardinia. It is absent or rare in the middle part of the peninsula around Rome.

Various ideas were entertained about the causes for this. Since it was common in some seaports it was thought to have been "brought in by sailors," navigators having long been known as transporters of genes to new localities. But other evidence did not indicate racial mixture as the cause and there were no obvious foci of the higher frequencies of the gene which would have been necessary to provide the donors or immigrants. But, when the distribution was compared with that of endemic falciparum malaria, the correspondence was striking: all high frequencies occur in areas formerly malarial.

A good example is the distribution of malaria and of thalassemia in Sardinia, as shown in table 3. Two neigh-

Table 3. The frequency of persons diagnosed by blood examination as having thalassemia minor (heterozygotes Th/th) in four Sardinian villages. (After Carcassi, Ceppellini, and Pitzus.)

Village [a]	Number of adults examined	Percent with thalassemia
Orosei	250	18.83
Galtelli	185	21.28
Desulo	320	3.75
Tonaro	107	4.67

[a] Orosei and Galtelli are lowland villages, about 4.5 km apart, in a sea level malarial zone. Desulo and Tonaro are mountain villages, about 8 km apart, at an altitude of 1000 m in a nonmalarial zone. The latter two villages are about 60 km from the former two.

boring villages at sea level and two at 1000 meters altitude were chosen. Although malaria was eradicated in Sardinia about 1948, it was formerly endemic in the lowland villages, absent in the mountains above the habitat level of the malarial mosquitoes. About 20 percent of the lowland villagers were thalassemia heterozygotes, but only about 4 percent of the mountain people. Statistically, the difference is highly significant. The Italian investigators were able to show that this difference was not due to racial difference between the lowland and mountain peoples. Blood group examinations in all the villages showed frequencies typical of Sardinia and unlike those of the mainland of Italy, the most likely source of immigrants with a high frequency of thalamessia. Although there is as yet no direct experimental evidence that thalassemia heterozygotes are more resistant to malaria, the ecological facts make this highly probable, and there is evidence that the homozygotes for this gene. those with Cooley's anemia, are resistant, for the Italian physician Ortolani tried in vain to infect such patients with malaria.

Moreover, there is evidence that among children born since 1948, when malaria was dramatically eliminated in Italy, there are no longer enough heterozygotes to maintain the gene at its former equilibrium values. Its frequency appears already to be falling and the state of balanced polymorphism (dependent on the selective advantage of heterozygotes in malarial areas) is probably being succeeded by one of transient polymorphism in which the frequency of the gene will approach zero. If this does happen we shall have a striking example of human interference with an evolutionary process, one in which the application of a good public health measure, the elimination of mosquitoes, had as one of its results the reduction of the suffering and fear caused by an "incurable" hereditary disease.

G6PD DEFICIENCY

I have already mentioned an interesting gene in the X-chromosome which in a mutant form interferes with the ability of the red cells to carry out certain steps in cellular respiration. Ordinarily males with the mutant gene in their single X-chromosome are not sick, nor are females who have it in one of their X-chromosomes. But when a person with the defective gene eats raw broad beans (*Vicia fava,* the fava bean) or inhales the pollen from flowers of this plant he (or she) may suffer a severe and often sudden anemia (favism) resulting from the destruction of many red cells. Similar crises may be precipitated in persons with the defective gene by ingestion of drugs such as the antimalarial primaquine. Some of the first information about this gene came from the study of Negro children who had eaten mothballs and become very ill. Naphthalene destroys red cells which have the defective gene. It appears that the distribution of this gene in human populations closely resembles that of thalassemia and probably for a similar reason, namely, that persons who carry it are more resistant to falciparum malaria. Thus its frequencies may be expected to be high in populations in malarial or formerly malarial areas. It is also subject to ·ther selective forces (such as the risk of favism) which tend to reduce its frequency.

When the distribution of this gene comes to be studied in relation to others found in the same populations, for example, sickle-cell anemia and thalassemia, we shall be in a position to make some interesting model⌐ of the operation of evolutionary forces on human populations.

IV

Genes and Evolution

The statement that no two humans are alike has come nearer to being proved by the variety of human blood groups than by any other specific hereditary characters. By "blood groups" we designate not merely the classic four groups of persons first recognized by Landsteiner in 1900 as differing in one property of their red blood cells, but all of the hereditary variety which has been proven to exist in those peculiarities of the red cells which cause those from certain persons to form clumps, to agglutinate, when placed in a specific testing fluid while those of other persons remain free, unagglutinated. The "clumpers" can be arbitrarily assigned to one blood group, the nonclumpers to another. There are thus as many groups discernible as there are specific reagents with which to test them. I myself belong to blood-group B because my cells whenever tested clump when placed in blood serum known as anti-B obtained from persons of blood-group A or O. I belong also to the blood-group "Rh-positive" because my cells also clump when placed in a specific serum designated as anti-Rh, or anti-D, whereas those of other persons, those of my wife, for example, do not clump in this serum. By similar criteria I know that I belong to blood-group Kell-positive, to the group MN, to Duffy-positive (named for the lady who first supplied the antiserum), and so on through a half dozen others for which my red cells have been tested.

The basis of each such agglutination reaction is the presence on the surface of the red cell of a chemical entity known as an *antigen* which shows its specificity by being able to combine with a reciprocally specific substance known as an *antibody*. When my red cells are placed in serum with the anti-B type of antibody, antibody molecules or groups of molecules attach themselves firmly to the surfaces of the cells, like keys fitting into locks; but since each antibody element bears several identical keys one of these may fit into a lock on one cell and another may attach to another cell, thus holding groups of cells together in a clump which can be seen with the naked eye or a low-power microscope. Each antigen is recognized by this lock-and-key reaction with a specific type of antibody. Each type of blood-group antigen thus revealed on the red cell owes its specificity to a gene of the individual from which the red cell was taken. These antigens are hereditary properties of the individual just as truly as the color of his eyes or the character of his hemoglobin. It is thus possible to describe that portion of the heredity of any individual which expresses itself in his red cell antigens by testing his cells in a variety of specific antiserums. A large number of different antiserums are now known, and the patterns of combination of the various antigens are so many that they seldom repeat each other in different individuals, descended as each person is from different ancestors who themselves had different patterns.

The antibodies are found in the fluid part of the blood, the serum. Just how they originate is still something of a mystery; but enough is known to permit them to be identified and to be deliberately produced for use. In fact, this step is the essential one in the whole process of blood typing, just as the preparation of reagents of known purity is an essential one in chemistry. The first antibodies of this type to be recognized were those which occur naturally and normally in the serum of individuals of the four classic blood groups. Thus, persons whose red cells are of type B (like mine) always have in their serum antibodies known as anti-A; persons of cell-type A have anti-B in their serum. If we take some serum from each of these two kinds of persons and put into it the washed red cells of some other persons chosen at random we find the types of reactions shown in Table 4.

Table 4. Reaction of cells to testing serums (+, agglutination; —, no agglutination).

Cells from persons	Anti-A	Anti-B	Cells belong to group
1	—	—	O
2	+	—	A
3	—	+	B
4	+	+	AB

All persons show one or another of these four reactions, that is, belong to one of these four blood groups. Group A persons always have anti-B in their serum; B persons have anti-A, O persons have both anti-A and anti-B, and AB persons have neither type of antibody.

INHERITANCE OF ABO BLOOD GROUPS

More than twenty years elapsed after the discovery of this classification before it was proved that the reactions of the red cells to anti-A and anti-B serums are determined exclusively by combinations of three alternative forms, three alleles, of one gene. Proof was obtained from two sets of observations, one on the inheritance pattern of these antigens within families, the other on their relative frequencies within a population.

From matings in which both parents are AB, children of three and only three types appear: A, AB, and B; type O children are not found in such families. When the offspring of a large number of such matings are added together, the three phenotypes occur in the proportions ¼ A, ½ AB, ¼ B. This is clearly the result to be expected from matings of two heterozygotes, and so we may begin by assuming an allele A responsible for antigen A and an allele B responsible for antigen B. Both genes express themselves in the heterozygote as in the MN case (Chapter II) so neither one is dominant to the other. Where both parents are group O, the children are always of group O. Group O persons may therefore be assumed to be homozygous (since they breed true) for an allele O. What now is the relation of O to A and $B?$ We look for matings in which one parent is AB, the other O, and find that all the children are either A or B, never O, just one half of each type. This tells us that O is recessive to A and B. We then examine matings of A $\times$ O and find that some A parents produce only A children; such parents may be homozygous $AA;$ other A parents produce both A and O children, half of each. Such parents must be heterozygous AO. Similarly some B parents are homozygous $BB,$ others are known to be BO because they are parents of children of blood group O. This makes six different genotypes accounting for the four phenotypes as follows:

Genotype	Blood group phenotype
AB	AB
AA	A
AO	A
BB	B
BO	B
OO	O

This means that there are three alternative forms that these elements, which show simple segregation from each other, can take—the alleles A, B, and O. These alleles were named before the convention came into general use by which alleles are always indicated by a symbol common to all. Now we should perhaps call these alleles, L^A, L^B, and l (L for Landsteiner).

Final proof that three alleles are concerned is obtained by computing the proportion of each of these in a population. This involves a little algebra and will not be worked out here. (Books on blood-group genetics such as those of Boyd, Mourant, or Race and Sanger can be consulted if desired.) When the allele proportions have been obtained for a given population, the proportions of phenotypes expected in that population can be computed from the Hardy-Weinberg formula on the assumption of random mating and the expected values compared with those actually found. Agreement between the predicted and the actual values indicates that the theory of random segregation of three alleles is correct.

This tells us something about the antigens on the red cells; each specific antigen is caused by a specific allele and we have to suppose that at some time in the past a gene mutated twice, either $A \longrightarrow B \longrightarrow O$ or the reverse, or $B \longrightarrow O \longrightarrow A$ or some other order. Such events are difficult to detect in human families. Children of group A or B born to presumed O parents could be due to mutation $O \longrightarrow A$ or $O \longrightarrow B$, or they could be due to illegitimacy or assignment of a baby to the wrong parents. All such cases thoroughly investigated by other means have been due to the latter causes. In spite of the fact that mutation in this gene may be rare and is hard to detect, it still seems reasonable to attribute the origin of allelic differences in this and other blood group systems to mutation. Similar allelic differentiation is found in other primates and it is possible that the mutations responsible occurred before the *Homo sapiens* stage of human evolution was reached.

Although we know something about the system by which the AB antigens are inherited, we know of no comparable system to explain the antibody differences, because segregation of genes governing antibody production has not been observed. One guess is that each individual is exposed, for example, to antigen A which occurs in a variety of animals, plants, and bacteria. To a child inheriting antigen B antigen A would be a foreign substance to which he might respond by making anti-A; to a child inheriting antigen A, it would not be foreign and therefore he would not respond. But there is no proof of this and we shall have to base our ideas for the present on the inheritance of antigens.

What led to the ABO classification in the first place was the discovery that transfusion of blood from one person to another sometimes led to shock or death of the person receiving the blood, that is, some bloods were incompatible with each other. Landsteiner found the reason for this incompatibility. When, for example, blood from a group A person is transfused to a B person, the latter contains antibodies against antigen A; the A cells are thus caused to agglutinate, the clumps tend to clog small capillaries in the lungs and elsewhere causing shock or sometimes death. As is now well known and in daily use in hospital transfusion services and blood banks, the compatible and incompatible combinations are as follows:

Donor	Compatible recipient	Incompatible with—
O	O, A, B, AB	None
A	A, AB	O, B
B	B, AB	O, A
AB	AB	O, A, B

Persons of group O are the universal donors; AB, the universal recipient, can receive blood from anyone.

It is also possible to make an anti-A serum by injecting red cells from an A person into a rabbit or other experimental animal. The animal responds to repeated injections as he would to any foreign proteins, such as those of a bacterium; he makes antibodies directed specifically against antigen A, that is, he becomes immunized and for a period thereafter his blood serum contains anti-A. Of course he may also make antibodies against the other antigens of the injected red cells, and, if a specific anti-A serum is wanted, these other antibodies will have to be removed by methods known to serologists, leaving anti-A in the purified reagent.

Antibodies specific for other human blood antigens arise occasionally by this same immunization process when an individual has received incompatible blood by transfusion or when a mother is immunized by carrying in her body a baby who has received an hereditary antigen from the father which is different from that of the mother. These two occurrences, especially the latter, have provided most of the specific antiserums newly discovered in the last 15 years.

THE RH BLOOD-GROUP SYSTEM

It was the recognition of active immunization of the mother by her own fetus, and the damage to the fetus caused by antibodies so produced by the mother, that led to the discovery of Rh-incompatibility. In 1939, Philip Levine, a former associate of Landsteiner, and his collaborator R. E. Stetson reported finding a new type of antibody in the blood serum of a woman who at her second pregnancy had delivered a stillborn fetus. The presence of antibodies in her blood was revealed when, following parturition, she was given a transfusion of blood from her husband. Both she and her husband were of blood-group O, that is, "compatible" as donor and recipient. Nevertheless, she suffered a severe transfusion reaction as though her husband's cells were incompatible with something in her serum. It was then shown that her serum would regularly agglutinate the red cells not only of her husband but also of some 80 percent of other blood donors of blood-group O. Obviously a new type of incompatibility had been found that was not due to the ABO system nor, as shown by other tests, to the other blood group differences then known, namely,

MN and P groups. The authors concluded that there was a new type of antigen, which the husband had but which the wife lacked. The fetus was thought to have inherited this antigen from the father. Entering the blood stream of the mother from the fetus, this antigen had evoked the formation of a specific antibody in her, just as though she had received a foreign antigen. She had been immunized by her own fetus. When her husband's blood was transfused, these antibodies reacted against the antigen in his red cells.

Shortly after this, in 1940, Landsteiner and Alexander Wiener identified a new antibody in the serum of rabbits and guinea pigs which had received injections of blood from rhesus monkeys. They called it anti-Rhesus (anti-Rh for short) and showed that it also agglutinated the red cells of about 85 percent of a large group of white blood donors from New York City, that is, about 85 percent of white New Yorkers were Rh-positive (had the Rh antigen) and about 15 percent were Rh-negative (lacked the Rh antigen). Soon this same anti-Rh antibody was found in the blood serums of some persons who had suffered transfusion reactions after receiving blood from donors of the proper or compatible ABO blood group. Finally, the antibody found by Levine and Stetson was shown to be the same as the anti-Rh discovered by Landsteiner and Wiener. It was, moreover, shown that Rh-positive had a dominant gene *Rh* leading to production of the Rh antigen; some were homozygous *Rh Rh,* others heterozygous, *Rh rh.* All Rh-negative people, those who lacked the antigen, were homozygous for the recessive allele *rh rh.*

A decisive fact was that the mothers of babies suffering and often dying from a severe hemolytic jaundice known as erythroblastosis foetalis were nearly always Rh-negative, the fathers Rh-positive, and the babies with the disease, Rh-positive also.

Now a general theory could be put forward based on

the work of Landsteiner, Levine, Wiener, and their collaborators and confirmed by much experience of others. It is outlined in the diagram in Fig. 7.

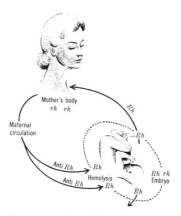

Fig. 7. The assumed relation between genes of the Rh blood-group system and hemolytic disease (erythroblastosis) of the newborn in a child from an Rh-negative mother and an Rh-positive father. The induction of anti-Rh antibodies in the fetus is assumed to occur in earlier pregnancies and to cause hemolysis of fetal blood cells in later pregnancies. (From Sinnott, Dunn, and Dobhzhansky, *Principles of Genetics*, 5th ed., by permission of the publishers, McGraw-Hill Book Co.)

The first step is the inheritance by the baby of the dominant *Rh* gene of the father. This leads to the production of Rh antigen in the red blood cells of the fetus. Somehow this antigen passes across the placenta, the network of blood vessels through which nourishment diffuses from mother to embryo. When the antigen gets into the mother's blood stream it acts like a foreign protein (remember she does not have the Rh antigen) and antibodies are formed, and her serum thus comes to contain specific anti-Rh antibodies. If this is her first

Rh-positive baby, these antibodies may not be produced soon enough in the embryonic life of the fetus to affect it; but if the formation of antibodies has been provoked by earlier pregnancies of this sort (or by transfusions of Rh-positive blood) so that she has become immunized, then the red cells of the Rh-positive fetus are attacked and hemolyzed; at birth the child has severe jaundice, is unable to oxygenate its blood and tissues, and may die shortly. Not all Rh-positive children of Rh-negative mothers are doomed to have this disease; some mothers do not produce antibodies and sometimes children are not fatally injured by them. In New York City at the time of this discovery about 1 baby in 200 suffered from this disease. A way of saving many of these babies by massive blood transfusions has now been found.

BLOOD GROUPS AND NATURAL SELECTION

But the essential fact is that, up until recently and where this therapy is not practiced, these deaths were selective, genetically, for it is always the Rh-positive babies which run the risk of death from this cause, and these are always heterozygous *Rh rh* (*Rh* from father, *rh* from mother). The Rh-negative babies escape the risk since they have no Rh-antigen to be attacked. Whenever a baby dies from this cause, therefore, two genes are eliminated from the population, an *Rh* gene and an *rh* gene. But since in populations of European descent *Rh* genes are more common than *rh* genes, the pool of *rh* genes is depleted faster than the *Rh* pool and *rh* should tend to disappear from the population. The fact that it has not been eliminated in certain populations indicates that other forces are acting on it which we must discover. What is important is that the ratio of *Rh* to *rh* is subject to change by natural selection. One of the causes of change in frequency of these Rh alleles has thus been discovered, namely selective death of certain genotypes due to parental blood incompatibility.

It must have been a cause of great satisfaction to Karl Landsteiner that, forty years after his discovery of the ABO blood groups, he and two of the students he had trained in his adopted country should open a new field, the relation of hereditary blood antigens to disease and thus to natural selection, which gave a major impetus to the study of human genetics and of the mechanism of human evolution.

The discovery of Rh opened the way to the discovery of many more blood factors. Some of these proved to be modifications (alleles) of the Rh gene itself of which now some eight forms are known: R^1, R^2, R^0, R^z, r', r'', r^y, r. The capital letters designate alleles which give the Rh-positive reaction, the small letters the Rh-negative. Other antigens known are independent of Rh. One of the latter is the Kell antigen due to a dominant gene which is uncommon in most populations, absent in some. Incompatibility for Kell is occasionally the cause of death of a baby from hemolytic disease, acting like Rh-incompatibility, and it is possible that Kell incompatibility causes other diseases. In the case of Kell, since it is the dominant gene which is less frequent, this is the one which is selectively eliminated by such deaths. Some of the other newly discovered hereditary antigens occasionally cause hemolytic disease in babies; for other antigens such effects are not known.

One peculiar fact about all the blood antigen genes is their varying frequencies in different populations. This suggests that there are relations between the effects of such genes and the conditions under which different populations have lived in the past, that is, that the frequencies of these genes have been influenced by natural selection. The investigation of this possibility began as soon as the genetic system responsible for the ABO blood groups and the differing proportions of these genes in different racial groups became known, but it is only recently that any clear positive results have been obtained connecting these blood groups with disease.

ABO BLOOD GROUPS AND DISEASE

Several cases of associations of a specific disease with one of the four blood groups have recently been found. The method of testing whether such an association exists is to compare the frequencies of the blood groups in a group of patients suffering from a disease with the frequencies in a comparable group of people drawn from the same population but not suffering from the disease. For example, in several British hospitals, comparison was made of the blood groups between patients with peptic ulcer and those hospitalized for other causes. The difference between these two sets of patients was striking: the ulcer patients contained a much higher proportion of persons of blood-group O, while the control group had only the proportion expected in the general population of that area. This was confirmed by a careful study of a large group of ulcer patients at a large Liverpool hospital that showed, in addition, that it was especially among patients with ulcer of the duodenum, the upper part of the small intestine, that the high proportion of blood-group O is found. The association of blood-group O and ulcers has now been found in Britain, Denmark, Switzerland, and the U.S.A. This means that it is probably not due, as was first supposed, to mixing of populations, one of which happened to have a high proportion of blood-group O and of people susceptible to ulcer, the other having lower proportions in both respects. It is unlikely that mixtures of just this kind would be found in several different populations.

The reasons for this striking association are in fact unknown although one other difference between ulcer and nonulcer patients may provide a clue. More of the latter than of the former were found to have a dominant gene known as the secretor gene, which determines that

the A and B antigens will be present not only on the red
blood cells but also in secretions such as gastric juice,
saliva, sweat, and tears. In persons with the recessive
allele for nonsecretion, the antigens are restricted to the
red blood cells. It may be that presence of the antigens
in the body fluids affords some protection against developing an ulcer. The risk of the disease, at any rate,
appears to be highest among group O nonsecretors,
lowest among group A or B secretors. Thus, at least one
gene other than those of the ABO system influences the
risk of getting the disease. The risk may also be influenced by environmental factors. We do not know, for
example, whether ulcers are more dangerous or less so
to people living in regions in which the gene O is rare
than in those in which it is common.

Let us take now the fact that in populations of
European descent persons of group O run a risk of
developing a detectable duodenal ulcer which is about
35 percent greater than persons of groups A, B, or AB.
Then let us try to determine what effect this would have
upon the frequency of these blood-group genes in their
descendants. Would the proportions of the genes change
because of this risk? The essential evolutionary question
is whether fewer O genes would be passed on in such a
population. This would happen if the number of children
of O parents that live to reproductive age were propor-
tionally less, because of the risk to the parents, than
those from A, B, or AB. There are no decisive data on
this point but at least we know what kind of data we
must have. We can guess that the effect will not be great
since ulcers usually develop at a mature age and excess
mortality caused by them will generally occur after some
children have been born and the genes passed on. Other
indirect effects are possible, of course. If persons who
develop ulcers differ in temperament from those who do
not, will they for this reason be discriminated against as
mates, be less fertile, more subject to other illnesses,

and so forth? There is little likelihood at present of separating effects due to such causes from the large number of other variables affecting family size.

Finally, we must consider the context, both social and biological, in which such events occur. There has been a persistent impression among physicians that ulcers of the digestive tract are associated with such aspects of modern civilized life as urbanization and industrialization, with consequent increases in "stress," whatever that may mean. However, ulcers also occur in nonurban, nonindustrial populations. It may be, of course, that ulcers get looked for in city people, and found because looked for, more often than under simpler conditions. Civilization has, however, other effects, one of which is that ulcers when found may be cured and the sufferer returned to the normal population. Moreover, if there were other genes, apart from ABO and secretor genes, causing ulcer more directly (and there is evidence of one such) the stirring up of the gene pool by random mating in cities would reduce the chance of these becoming homozygous.

Here again it is the way of thinking about causes of change in gene frequency that I have stressed, using the facts that are now beginning to appear as illustrative rather than decisive for detailed theories. There are signs that studies of blood-group genes in relation to disease and other forms of natural selection will now make more rapid progress. Already there is evidence of an excess of blood-group A in patients with gastric carcinoma (stomach cancer) or with diabetes, and there are hints concerning other diseases. A most important lead has been opened by the discovery that women of blood-group O married to men of group A or B have fewer living children than A or B women married to O men; and incompatibility of mother and fetus in the ABO genes is known to be a cause of hemolytic disease of the newborn. Mothers of blood-group O have, of course, both anti-A and anti-B in their blood (and this is potentially dangerous to their unborn A and B children) and it is likely that this and other immune reactions associated with ABO account for some portion of the early abortions and miscarriages which are still common even under good medical care. Deaths of unborn children, if selective in respect of genotype, would, of course, be the most potent of all effects in changing the gene frequency in subsequent generations. It is likely that strong effects of this kind do operate on the frequencies of the ABO genes to a greater extent even than in the case of Rh, since the proportions of ABO are notoriously more variable than those of the Rh alleles in closely related neighboring populations. There has been interesting speculation that the great plagues of the past and some infectious diseases today have tended to leave more survivors of certain blood groups than of others. This is the kind of massive influence of natural selection which would be effective in changing the frequencies of blood-group genes; but such effects have not been proven.

It is probable that natural selection acts not only to alter the frequency of genes of the ABO and Rh systems in the above ways, but also on specific combinations of ABO and Rh genes. For example, fewer unborn Rh-positive babies get erythroblastosis when the Rh-negative mother is blood group O and the father is both AB and Rh-positive. The O mother's blood contains antibodies against both A and B cells, and if any of these from the fetus should penetrate the placenta, they would probably be destroyed. Since in this case they would also be Rh-positive, the red cells of the fetus would thus be prevented from provoking the formation of antibodies against Rh and it is these antibodies from the mother which are responsible for Rh erythroblastosis. The frequencies of ABO and Rh alleles probably depend to some extent on each other in this rather paradoxical effect of selection.

The most general view we can reach at present about the arrangements of the blood-group genes in human populations is that they exist in a state of balanced polymorphism in which the equilibrium is maintained by selective forces of various kinds of which at present we have only a preliminary and incomplete listing. The most direct way of detecting elements in this equilibrium and changes in it would be to compare the frequencies of different blood-group genes in the same population of people at different ages. As the population at birth, or ideally at conception, is reduced by mortality, is the reduction selective in respect of any of the phenotypes? People can now be described in terms of some ten blood-group systems (ABO, MN, P, Rh, Kell, Duffy, Lewis, Lutheran, Kidd, secretor), some of them accounting for many different phenotypes because of the large number of allelic alterations known in some of these genes. A very large-scale study begun now should reveal within one generation whether different phenotypes have different life expectancies or different risks from particular causes of death. Insurance companies would eventually profit from such knowledge if it were to permit them greater accuracy in computing risks.

Another way to study the same question of selective survival is to count the numbers of genes of a given kind in a large population of parents and the numbers of these in their children at a given age. This requires diagnosis of genotypes and not merely of phenotypes, but where phenotypes of both parents and their children are known it is often possible to deduce the genotypes. For example, a child of blood-group O from parents both of whom are of blood-group A reveals at once that both parents have an O gene, both are heterozygotes AO; and similarly with other genes which show dominance.

Eventually this sort of thing will have to be done on a large scale to study one of the most important questions of all, whether heterozygotes enjoy some selective advantage over homozygotes, as in the case of the sickle-cell gene in a malarial environment. The same is probably true of the gene for thalassemia and possibly also for the G6PD enzyme deficiency (cf. p. 64). There is a good indication that persons of the MN blood group, who are of course heterozygotes, are more fit in an evolutionary sense than either of the homozygous types MN and NN. It is quite possible that it is this superiority of the heterozygotes that is responsible for the retention of all three genotypes in most populations in a state of balanced polymorphism. One of the striking facts revealed by studies of human genetics, and especially of the blood groups, is that most of us are heterozygotes, usually for many different genes. There must be some evolutionary sense in this; is it that populations which are variable have greater plasticity, greater ability to adapt to a greater variety of environments?

RARE AND DELETERIOUS GENES

Before concluding this incomplete survey of facts and possibilities suggesting causes of changes in gene frequency, I must give some account of those human genes which until recently occupied the forefront of attention in books about human heredity. Many people gained the impression that human genes that could be identified and studied were mostly those with bizarre and abnormal effects: six fingers rather than five, albinism rather than normal pigment, idiocies, crippling disabilities, color blindness, bleeder's disease. The roster of what popular writers have called the black genes or "the bad ones" is long and dreary and one might well ask after reading this list: are not our good qualities inherited? They are, to be sure, but usually many genes are involved in these, and such cases are notoriously difficult to analyze even in animals and plants which can be studied experimentally. The qualities which give use and value to crop plants or agricultural animals are generally those like size, yield, reproductive efficiency, quality of proteins or of other determinants of food value which are affected by complexes of genes with individually small effects. These are known collectively as polygenes or multiple factors. Since they can usually not be recognized by the ordinary methods of genetics, they have to be dealt with by special statistical methods of analysis and this leads many investigators to avoid them. In spite of the fact that genes of this sort are probably more important with respect to natural selection and thus in evolution, we know much less about them. It is probably the "good" genes of this sort which have enabled the great bulk of mankind to cope successfully with past and present environments—otherwise we should not be here as a rapidly expanding species.

It was a natural and inevitable sequence that bad genes should be noted and studied first. In the first place, they are easy to spot. Surely a sixth finger is more apparent than a five or ten point difference in I.Q.; it appears at birth, is unaffected by diet, schooling, and all the other variables that affect intellectual development. A gene is identified only when it conforms to the rule of segregation and this can be most easily proved where the difference caused by the two forms of the gene allows accurate classification into sharply distinct categories. Attention is, of course, first directed to these, especially the attention of physicians who often make the first reports of abnormalities which appear several times in the same family. The interest which prompts a man to write a paper for a medical journal on a case of congenital heart defect would hardly lead him to write a communication on a heart which functioned with peculiar excellence, even if he could identify it.

The studies of our sharper shortcomings served the useful purpose of proving that Mendelian heredity operates in man just as it did in Mendel's peas; and it may be well to recall that that first demonstration encountered the opposite criticism, namely, that genes influenced only such superficial and unimportant characters as flower color or seed shape, which were said to have no importance in evolution. Both of the criticisms, that genes produce disastrous effects or merely superficial ones, we now know to be beside the point. The important thing is that genes provide the particulate method of inheritance the results of which are sifted by the environment to produce the adaptedness of the population.

Looked at in this way, the genes that produce effects which appear to be nonadaptive under any known environment, such as hemophilia or crippling congenital abnormalities, may be viewed as the price we pay for the primary advantage of having genes which can change by mutation. Such genes, in spite of the space they occupy in the literature of human genetics, are individually all quite rare in any population. Most of them are probably the results of relatively recent mutations.

Let us take the case of hemophilia or bleeder's disease, since this also illustrates the method of sex-linked inheritance. This was the first human disease to be identified as due to a gene which travels in the X chromosome from mothers to sons. In a person with hemophilia the blood fails to clot promptly enough to stanch a wound or to prevent internal bleeding, into the joints, for example. As a consequence most hemophiliacs die before leaving offspring. There are now known a number of different hereditary forms of hemophilia, all rare. In what follows I shall speak only of the first-described "classic" hemophilia. This occurs almost exclusively in males. In Denmark, for example, about 13 males per 100,000 born have this disease. The inheritance of it is clear and simple. From mothers who are heterozygous for this gene married to normal men, about half the sons inherit the disease, the other half are free from it and do not transmit it. The daughters do not show the disease, but half of them, in turn, transmit the gene to half of their sons. The gene follows exactly the distribution to the gametes and offspring of one of the chromosomes, the X chromosome, of which females have two, a pair, but males only one. Each egg, therefore, has one X along with one of each of the 22 other pairs of chromosomes, which are known collectively as autosomes. Sperms, however, are of two kinds: half get an

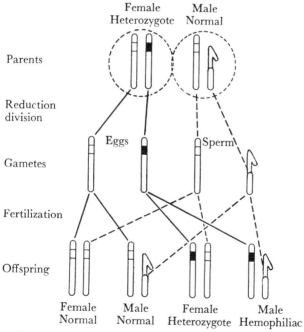

Fig. 8. The transmission of a gene for hemophilia by the X (sex chromosome) from a heterozygous (carrier) mother to half of her sons.

X at the reduction division, half get an unequal partner of the X, known as a Y chromosome, which occurs only in males. The course of inheritance of these chromosomes and of a hemophilia gene carried by an X chromosome is shown in Fig. 8. An essential assumption is that the Y chromosome has no representative of this gene, neither the hemophilia gene nor its normal allele. Consequently, a male that gets the hemophilia gene in the X chromosome from the mother cannot get the normal allele. Females, however, do receive from a normal father a normal gene in the X chromosome and do not show the effects of a hemophilia gene received from the mother. Since no constant difference between "carrier" females (heterozygotes) and "noncarriers" (homozygous normals) has been established (although there are hints that a difference may yet be proved), we refer to the hemophilia gene as recessive.

This means that every male that gets the gene for hemophilia shows its effects. Since one of its effects is to make it likely that death will occur before the age of reproduction, the gene is subject to strong selection against it and should decline quickly in frequency in the population. It is estimated that a hemophilia gene newly arisen by mutation has an average life expectancy in the population of about three generations, since about a third of the genes descended from it are eliminated in each generation. On this basis it can be calculated that, to account for the proportions of this hemophilia gene in a European population at any time, new mutation from the normal form to the recessive allele must be occurring at about the rate of 1 gene in 50,000. We might say, therefore, that such a gene cannot have a lasting effect on the adaptive powers of the population except to lower it slightly. The frequency of the gene is probably due primarily to its mutation rate, which means it will always be rare, unless something occurs to increase its mutation rate, or unless heterozygotes for it have some advantage, at present unknown.

Again, however, we must assess the prospective effect of such a gene by reference to the context in which it occurs. In a human society like that of nineteenth-century Europe, with a royal caste, much might depend on just who got the gene, rare though it was. Queen Victoria happened to be a carrier who transmitted it to some of her sons and to the Russian and Spanish royal families with whom her daughters intermarried. The course of European history may well have been altered by this gene. If a complete cure for hemophilia is found it may increase; if carrier females can be identified and persuaded not to have children, the life span of a new hemophilia mutation may be still further reduced. It is unlikely that human populations will ever be free of it, because of the likelihood of occurrence of new mutations.

An interesting case of a mutant gene which has become deleterious under modern conditions has been worked out by Geoffrey Dean and others in South Africa. They found that about one percent of the descendants of Dutch settlers there carry a dominant gene for porphyria. Such persons are detected because they excrete in urine and feces abnormally large quantities of porphyrin, a red pigment arising from the breakdown of hemoglobin. They are not sick and ordinarily suffer only minor inconvenience. But when they are given barbiturates and certain other drugs they may be poisoned, and some have died. This gene entered South Africa in a Dutchman who settled there in 1686, and by chance spread in the Afrikaner population since, before barbiturates were introduced, it did not lower the fitness of its possessors.

We might discuss other genes with very deleterious effects in similar terms. Many of those heretofore identified are dominants, like epiloia, a fatal abnormality of the brain of which all cases known are probably due each to a separate mutation with a rate of about 1 per 100,000 genes; or chondrodystrophic dwarfism, with a mutation rate of 1 or 2 per 50,000; a fatal eye tumor, retinoblastoma, with a mutation rate of about 1 per 50,000. These keep themselves rare by their fatal effects and low mutation rates. A well-known dominant gene is that for Huntington's chorea (St. Vitus's dance), which is somewhat more frequent. Because of the late onset of symptoms, usually after part of the reproductive age has been passed, natural selection is less effective in keeping this disease in check. A fatal disease, Kuru, expressing itself in progressive paralysis, is known only amongst a small group of natives in New Guinea, in which half of the females and a tenth of the males who get the disease die within a few months. It is probably the result of a specific genotype, perhaps a single gene which by chance persists in only one area.

Most deleterious genes with recessive effects are individually rare. Take the hereditary mental defects which we are sometimes told (without good evidence) are increasing to a degree which endangers the progress of society. Identified genes with such effects are juvenile amaurotic idiocy, fatal in early life with a homozygote frequency of about 1 per 40,000 (gene frequency .005); Tay-Sachs idiocy fatal to infants, with somewhat lower frequency and occurring mainly in descendants of an East European Jewish population in which a recent mutation had probably occurred; phenylketonuric imbecility, also rare, in which homozygotes have a metabolic defect in the biochemical system which deals with an essential amino acid, phenylalanine, which is not utilized effectively in homozygotes and is utilized more slowly than normal in heterozygous carriers. In the former, permanent brain damage is caused; in the latter, apparently none. Microcephalic idiots, with very small brains, have another rare gene, and there are several others, all rare, with secondary effects on intellectual capacity. When idiocies (defects with mental age up to five years) and imbecilities (mental age to ten or twelve years) from all causes are added together they amount to about 1 percent of all births in European populations. The equilibrium frequencies of each of these genes are set mainly by the mutation frequency of separate and distinct normal genes. Although prohibition of cousin marriages, which brings them to expression (where two rare recessive genes are required they usually come from the same source in an ancestor common to both parents) might reduce the frequency of affected persons somewhat, some loss of fitness due to such genes is probably inevitable in human populations. The authority on mutation, H. J. Muller, has referred to this background rate of occurrence of deleterious mutations as "our load of mutations," which I shall refer to further in the last chapter. At present one can take it as the handicap with which any population starts its race with evolutionary destiny. It means that neither the starting point nor the goal of that race is biological perfection but adaptedness. Since the handicap is more or less a constant feature of human biology, we may be interested in how human populations have adjusted their biological structures to accommodate it, but we need not be further concerned with explaining changes in the frequency of such genes. Genes which are common and occur in different proportions in different populations are more likely to provide the clues we seek in the study of adaptedness. It must be evident, also, that on statistical grounds alone, rare genes would not be very helpful in comparing populations living under different conditions. If a disease such as amaurotic idiocy is found in some well-studied populations like that of Sweden only once in 40,000 births, we should have an even chance of missing it if we examined 40,000 people from another population in which it occurred at the same rate. Rare things are thus subject to great fluctuations by chance, a phenomenon known to statisticians as sampling error, and reliable estimates of frequencies, which are necessary in comparing populations, are not likely to be obtained for rare genes.

V

Race Formation

There can be little doubt that a prime factor in man's conquest of this planet was the formation of local populations differing in genetic constitution and adapted to local conditions. Any species which attains wide distribution does so by its ability to exploit the opportunities provided by a varied environment. This ability depends, in essence, on the possibility of assembling assortments of genes which differ from environment to environment and on maintaining a rough kind of equilibrium for the long period which the trial and error method of evolution requires for operation. The possibility is provided by the Mendelian method of heredity, which had evolved long before man appeared upon the scene, and by ways of using it in populations, such as balanced polymorphism, which had developed out of the historical necessities that faced our antecedents.

Race, in popular usage, is a word with many shades of meaning and connotations and has become so emotionally loaded that some scientists would like to do away with it altogether in referring to human groups. But to the evolutionary biologist it has a clear and unambiguous meaning. I shall use it in this sense: a race is a population which differs from other populations in the frequency of some of its genes (a population has already been defined, for crossbreeding species, as a community of genes shared by interbreeding within the group). The biological concept of race is, thus, a flexible and relative rather than a fixed and absolute one, and this is required by the function it serves in evolutionary thinking. Its function is to identify a stage or stages in the evolutionary process. Of course, it may serve a taxonomic function also in enabling us to sort out and put in some order the variety we find in any collection of organisms. But we put things in pigeonholes, in studying evolution at least, in an effort, not only to attain the order of tidiness, but to try to understand the pattern of the whole variety. We put together, as members of a race, populations which have many, perhaps most, of their genes in common. We want, not order for its own sake, but order for the sake of tracing relationships and evolutionary descent. So, for our purposes, the use of race as a classifying device is secondary to and determined by its use in helping us to unravel the story of how man attained his present variety.

It is obvious that, since race refers to the variety within a group within a species, we shall not always be able to classify single individuals by race. A British biologist, Sir Julian Huxley, illustrated the hopelessness of such an ideal and of the concept of race on which it is based by a rather mean trick he played on fellow biologists and laymen alike. In the frontispiece of a book on race (*We Europeans,* 1945), he reproduced the photographs of 16 Europeans, each a native of a different nation. The game was to guess where each of the 16 belonged. If you assigned the Frenchman to France, the Finn to Finland, and so on through the 16, you got a score of 100. I left that book open at this frontispiece on my living room table for some weeks and watched my visitors tackle the game with the air of superiority of well informed people who knew the races of Europe. But the high score was 50. Considering that you would get a score of about 6 percent correct fits by juggling 16 pictures at random into 16 numbered slots, that gave a maximum accuracy of 44 percent to "racial classification" by facial features. Of course it was unfair, but it had an important lesson: not all people of one racial stock are alike. Racial "types" or national types are abstractions which form in the mind but do not exist in nature. Huxley's point was that the sooner we get over thinking of race in that way the better for us.

We can, however, classify *groups* of people into races in various ways. We can assign all black-skinned people to a "black" race and all white-skinned ones to a "white" race; and deal with yellows and reds in the same way. We get into trouble with the browns because, in other ways, some seem to belong to the blacks, others to the yellows or whites. And within the whites we find swarthy skins and very pale skins, sometimes even within the same family. So we turn to other descriptive characters: hair form, nose form, head shape, and other bodily proportions; and, by using enough of these judiciously, physical anthropologists can classify quite well the existing varieties of men, but, of course, only by groups. Statements can be made, with various degrees of probability, about the provenance of single individuals but this is definitely a job for the expert. Such a classification of groups can result in a few races (like the great continental divisions of black, white, yellow, and red, which are generally agreed upon) or in many, depending on the problems which the classification is to elucidate. This is not a matter of prime importance except to troubled anthropologists who get called into court (either of law or science) to assign persons or groups to categories that often have been set up for nonbiological purposes. "Is this man a Negro?" more often asks for assignment to a social category than to a biological one, at least in America and South Africa.

Serviceable though the racial classification method of physical anthropology is when used by experts, it has one prime shortcoming for the student of evolution: it can tell us almost nothing about the frequencies within the different populations of the genes associated with physical characters, simply because we cannot deduce the genotypes from the phenotypes of the metrical characters which have been traditionally used. These, like head form, for example, depend on the interactions of many genes and are influenced by the physical environment and customs of different peoples.

Consequently, some anthropologists, about a generation ago, began to supplement their standard descriptions with others that revealed gene frequencies. The use of blood-group genes for this purpose was begun by the Hirzfelds, two Polish physicians, who noted among the allied forces assembled near Salonika in 1914–15 wide and characteristic racial differences in the proportions of the ABO blood groups. The method was extended to other peoples in different parts of the world, to other blood-group systems as they were discovered, and to other genes, such as those for sickling, thalassemia, taste blindness, color blindness, and a few others. The world literature on "The Distribution of the Human Blood Groups" was reviewed in 1954 in a book with the same title by Dr. A. E. Mourant, director of the British Medical Research Council's Blood Group Reference Laboratory and adviser to the Nuffield Blood Group Center of the Royal Anthropological Institute, two bureaus in London in which the results of blood typing from various populations are checked and collated. The cooperation and collaboration of these bureaus with investigators in all parts of the world has been an important factor in producing the world picture which is now beginning to appear.

The general situation revealed by this review encourages the belief that, in the multiplicity of blood-group genes and of certain others, an instrument is at hand by which some of the agencies involved in race formation can be profitably explored. We should recall, before looking at the evidence, some of the assumptions upon which this use of blood groups is based. An important one is that the phenotypes revealed by reactions of blood cells to antiserums are effects of genes more directly and immediately related to the characters than is the case with genes affecting morphological traits. The blood-group antigens generally appear before birth and the production of them is not known to be influenced by diet or other environmental factors. They are reliable indicators of genotype. Second, it is assumed that the same blood-group loci are present in the chromosomes of every population, the different forms (alleles) which the locus has assumed having arisen by mutation. In the case of any one blood group, it is the distribution of different alleles of the same basic element which is being compared between different populations. This assumes, essentially, that all varieties of man have shared a common pool of genes, that mankind is one species.

The results of the world survey of blood-group genes appear to justify both of these latter assumptions. Some form of the gene that determines the reactions of red blood cells to a specific antiserum appears to be present in every race which has been tested, and the same applies to the other blood-group genes. Human red cells are comparable from whatever source or environment they are obtained. There is thus good assurance that the same properties are being measured.

In general, the alleles of every blood-group gene for which adequate information is available show geographic variations in frequency; none is constant or uniform throughout mankind. Variation is the rule, not only in the distribution of variants of a single gene, but also in the combinations in which these appear. We have already seen that the variety of combinations is so great under the random mating prevailing even within one social community that hardly any two individuals are likely to have the same assortment of alleles. What we are to look for in the world distributions of genes, therefore, is not uniformity within any geographic area, but patterns formed by the arrangement of the total variety. We look, in other words, for significant differences in *proportions* of alleles between different groups of people, and measure the geographic extent of the group by the area over which similar proportions obtain. This, of course, is the method used by physical anthropologists for anatomical characters, with the important difference that now we use genes rather than phenotypes or "descriptive traits."

The first result of this kind of survey is to confirm in a general way the conclusions about major racial differences which had been reached by ethnographers. The great continental races—Europeans, Asiatics, Africans, Australians, American Indians—show the widest differences from each other in the frequencies of certain genes (Fig. 9) and in the characteristic combinations. Europe, for example, is the only continent in which the native inhabitants have the Rh-negative gene in high frequencies; elsewhere it is less common or absent. In Europe, however, the gene pool has been stirred up so frequently by migrations and by invasions from Asia and North Africa that few other generalizations can be made about so complex an area. In Africa, south of the Sahara, all populations tested have very high proportions (50 to 90 percent) of one Rh-positive allele (R^o), clearly and discontinuously different from all other populations, in which this allele is rare or absent. It continues to have this high frequency in populations of African origin who have been settled for many generations in other parts of the world, such as Negro communities in the New World. Serological ethnographers thus tend to refer to it as the "African" allele.

American Indians are unique in several respects, lacking the blood-group allele *B*, the Rh-negative allele *r,* the Kell (*K*) allele, and probably the nonsecretor allele; and having very high frequencies of *M*. The Diego-positive alelle, *Di*ᵃ, is present in variable propor-

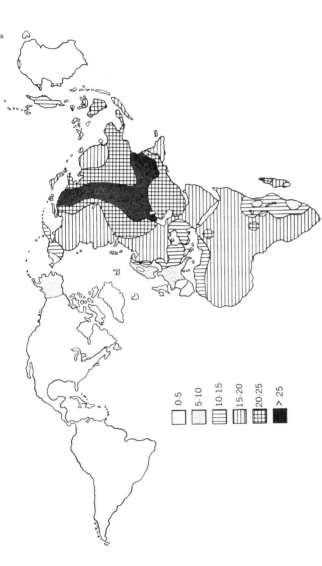

Fig. 9. The frequency, in percent, of the blood-group gene *B* in indigenous populations of the world. (From Sinnott, Dunn, and Dobzhansky, *Principles of Genetics*, 5th ed., by permission of the publishers, McGraw-Hill Book Co.)

0·5
5·10
10·15
15·20
20·25
> 25

tions in the few American Indian populations which have been tested for this antigen.

Asia, like Europe, has seen continued movements of peoples and, in addition to being racially complex, is also very incompletely covered by blood-group surveys. It contains the regions of highest frequency of the allele *B*, the Rh-negative allele is rare or absent in the populations of the Far East, and the Diego factor is common in Chinese and Japanese.

The Australian aborigines are as clearly marked off from other continental peoples by the peculiar frequencies of their blood-group genes as they are by physical and cultural characters. They lack the *B* allele, the *S* allele of the MNS system, and the Rh-negative allele, but have very high frequencies of *M*.

These descriptions are useful in establishing norms for key areas which have served as centers of dispersion of populations that now occupy intermediate areas such as Melanesia, Polynesia, and island groups generally. They pose some interesting problems concerning the genetic relations between the continental peoples themselves. Why, for example, should the allele *B* be absent in the American Indians, who came, according to the best evidence, from Asia, the center of the highest *B* frequency in the world? Invaders of Europe from Asia within historic times probably carried with them many *B* alleles, for the frequency of *B* declines quite regularly across Europe from east to west along the route of the invasions which thinned out as the distance from the points of origin in central Asia increased (Fig. 10). But why, if this was one of the causes of the present distribution of *B* in Europe, did this allele establish itself there but not in America? The relations between Asia and Africa, areas in which the most ancient evidences of man's origins have been found, are especially interesting but almost wholly unexplored by this method. The relative frequencies of the sickle gene in North Africa

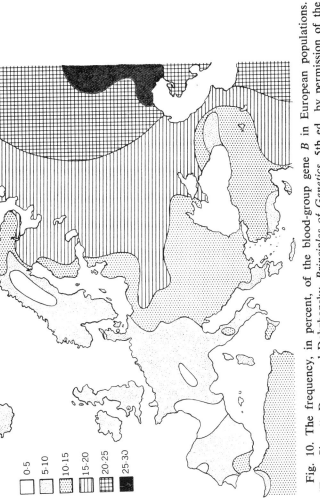

Fig. 10. The frequency, in percent, of the blood-group gene *B* in European populations. (From Sinnott, Dunn, and Dobzhansky, *Principles of Genetics*, 5th ed., by permission of the publishers, McGraw-Hill Book Co.)

0·5

5-10

10-15

15-20

20-25

25-30

and southern India can hardly be used as indicators of relationship in view of the convergent effects produced by similar (malarial) environments on the frequencies of this gene.

But it is the relations between smaller groups of people, within continental areas and between them, that the new methods are most competent to illuminate. Here, as the total problem of race formation is broken down into smaller and smaller parts, one is most likely to obtain clues to the causes of the differences in gene frequencies, and this, rather than description for its own sake, is the essential question for us. When we look at the mosaic formed by mapping the frequencies of the alleles of one gene we do get some clues. We are struck at once by the wide variation in the frequencies of the ABO alleles between closely contiguous populations that we know from other evidence to be similar and closely related. Detailed maps of these frequencies, based on hundreds of thousands of individual tests, exist for Britain, Switzerland, and northern Sweden, for many American Indian tribes, and for Australian aborigines, to choose several different sets of environments. In all of these we discover an order relating change in gene frequency to physical distance. As we proceed from north to south in Britain the frequency of blood-group O regularly declines; in north Sweden it is blood-group B which shows this regularity of change. In Australia it is gene *A* which increases from north to south. This change with distance of a descriptive trait or a gene frequency in a plant or animal population is known as a cline. Clines are important indicators of some aspect of the environment which is changing. Sometimes it is distance from another population which serves as the source of supply, through intermarriage, of the allele whose frequency declines with distance. A cline on a larger scale was that of decrease of B in Europe with increasing distance from Central Asia. An expression now in common use with geneticists is "gene flow" between populations. (They move, of course, in people who move and marry at a distance from home base.) A cline may be an expression of the direction of gene flow. Or it may be the expression of some natural feature of the environment, such as the cline (decline in this case) in the frequency of thalassemia with increasing altitude.

The slope of the cline may be suddenly changed by some natural barrier, as at the river Tweed in northeast England: the ABO frequencies on the north and south banks of the river differ more than they do between more distant points not interrupted by a river.

Other peculiarities in the distribution of ABO frequencies concern the increase in group O in marginal areas like Ireland and Iceland, and on a larger scale throughout the Indians of South America, who have very few A or B genes. South America was presumably peopled by migrants from the north.

But there are also sudden changes in the frequencies of these genes between nearby communities which have similar frequencies of other genes. The inhabitants of the two mountain villages in Sardinia (Table 3) differed significantly in ABO proportions, and so did the two nearby lowland villages, although all four had similar frequencies of MN and Rh alleles.

All of these facts suggest that the ABO blood-group genes are sensitive to evolutionary forces—but to which ones?

Differential mutation rates are probably not the cause, for reasons given above. Mixture between racial groups with different frequencies of these genes is certainly involved in many cases. Sometimes it is a profitable exercise to imagine what the gene proportions would have to have been in the hypothetical populations which may have contributed to a mixture. This kind of thinking suggested that some of the peculiarities of western European populations (low proportions of B, high proportions of Rh-negative) might have been the result of mixture between an indigenous population, with no or very little B and much higher proportions of Rh-negative, with invaders from the east with higher B and little or no Rh-negative. In casting about for a population to fit the description of the hypothetical "proto-European," the present-day Basques presented themselves. Here is a population with the lowest B in Europe and the highest Rh-negative frequency in the world, speaking a non-Indo-European language, and with some skeletal features resembling those of the men of late paleolithic times who left their skeletons and their tools in the river valley caves of western Europe. No wonder that some anthropologists regard the Basques as living witnesses of an ancient European race that elsewhere vanished into a mixture.

But all this is a little beside the main point, for what produced the greater gene frequency differences between the hypothetical races which entered the mixture? Race mixture is certainly a constant accompaniment of the later stages of the peopling of this planet, but it always leaves the causes of the earlier differences unexplained.

Here, two evolutionary forces have to be invoked for which some theoretical models can be deduced from facts about present-day populations. One of these is the accidental chance fluctuations of gene frequencies in very small populations. I have referred to this earlier as random genetic drift. Small populations are notoriously peculiar, often deviating in their gene frequencies from neighboring populations from which they are reproductively more or less isolated. I shall discuss some of these in a separate chapter on small communities. For a present instance to make the point clear, imagine a population in which 90 percent of the people are blood-group O, 10 percent blood-group A. A small band of migrants leaves the population and settles a new uninhabited territory, say, on the other side of a mountain range. For each member of the band the chance is nine to one that he or she will be blood-group O. There is thus a good chance that the gene A will not be represented in the migrants. If a new isolated community is descended entirely from such migrants it will be 100 percent group O; A will have been lost. Something like this may have happened on the west coast of South America where Indian communities on the two sides of the Andes differ in the above way.

It is inevitable that such accidents should happen in populations descended through small bottlenecks, but this is no proof that differences arose in this way, although until recently it was a view commonly held about the origin of population differences in blood-group gene frequencies. If this were to be considered the sole cause, it would tacitly assume that the blood-group genes are selectively neutral, that, to put it baldly, group A is as good on one side of the mountain as on the other. We have no assurance that the latter is true; in fact, suspicion that it may not be so grows with every new case of connection between a blood-group gene and a selective agency such as disease.

So we turn to the last evolutionary force— selection —which we may suspect from observations on populations of animals and plants and from such human cases as those of sickling and thalassemia to be a determining influence. Whatever other agencies may initiate changes in frequency, the new arrangement must be at least acceptable, compatible with the circumstances in which it finds itself.

With respect to the action of natural selection on the distribution of the ABO groups, we are simply not in a position now to cite positive or decisive evidence on which general statements could be based. Connections between blood-group genes and diseases of maturity like gastric carcinoma or duodenal ulcer, interesting and important as they are, do not in themselves provide decisive evolutionary effects. The fact that most large and successful populations, like those of Europe, Africa, and Asia, contain all four groups indicates that no one is clearly superior under all conditions. The very widespread occurrence of polymorphism in this and in most other blood-group systems indicates that it is some kind of balance which is adaptive in particular circumstances. It is probably balance not only within but between different blood-group systems which is important. We may recall the fact that incompatibility between parents for ABO blood groups actually protects from hemolytic disease the children of parents incompatibly mated for Rh. It is in populations with Rh-negative genes that the risk of loss of such genes occurs through the death of heterozygous babies; if such a population contains a variety of ABO blood groups, the risk of loss is reduced. One polymorphism tends to support the other.

Race formation, as viewed from the evidence of blood-group distribution, is thus the process by which coadaptive complexes of genes are assembled. The diversification begins by the origination of variety through mutation and the formation of all possible combinations of genes which is the inevitable consequence of bisexual reproduction. The frequencies of individual variants are subject to statistical fluctuation and are probably sifted by selective forces, largely unknown at present, acting upon complexes of genes to produce a balance compatible with each other and with environmental circumstances; the complexes may be sheltered for a time by reproductive isolation, but are subject to change through migrations and race mixture.

What now of the idea of race itself in this sort of evolutionary picture? Races endure for longer or shorter periods of time, but they eventually change as the world grows older. It is clear that there can be no single prescription for a successful or superior race. Purity, fixity, uniformity never seem to be goals in human evolution. What little approach to them we observe is in more or less isolated pockets here and there over the earth, where peculiarities are preserved for a time as if to offer to those who are curious about such things little glimpses of what we like to imagine was a simpler state in which our ancestors lived in small bands, small villages, small city-states, before size and complexity overwhelmed us. We must devote a chapter to an excursion among these small communities because they give us some idea of the part played by numbers of people and by isolating factors in forming population structure.

VI

Isolated Populations and Small Communities

One of the causes of change in gene frequency is the rate of gene-flow or migration between populations with different gene frequencies. It is measured by the degree to which a population fails to share its genes with other populations, that is, by the extent of its *isolation* from other populations by geographic or social factors which act as barriers to intermarriage.

The second of these causes of change has been referred to in Chapter III as random genetic *drift*, which may alter gene frequencies when the effective breeding size of the population, that is, the number of individuals who contribute gametes to the next generation, is very small.

Before we consider some examples of the effects of variations in these two factors of isolation and of size, we had better recall the sense in which we use the word "population" and the purpose for which we employ it. Since mankind is a single species, we shall use "population" in a restricted biological sense as denoting a group of people who share a common pool of genes through reproductive communication by intermarriage. Such a group may have no sharp boundaries; membership in it would have to be determined by the probability of finding a mate and passing on genes within the group rather than outside of it. The size of such a group is thus given by the distribution of these probabilities. For all of us there is a probability—albeit a low one—

of mating and having offspring with any other human being of opposite sex. The group included within these probabilities is as large as the whole species, and we may therefore think of the whole human race as one community of genes. There is a much higher probability that we shall marry within our tribe or local community; in fact, the probabilities are always highest for mating within some restricted circle, and from this core or nucleus the probabilities diminish as the circle widens.

A population in this biological sense has another dimension—duration—for, if the feature which interests us is the pool of genes shared by intermarriage, continuity is of its essence. All of us belong to communities, usually to several different communities at once. These will have biological significance for descendants only if membership in one or more of them has a long-continued influence on the marriage pattern. Membership in a long-established religious community often has this effect; sometimes it will restrict our choice to one of relatively few persons if the community is small, like that of the Dunker community or the small Jewish community described later in this chapter. Or our choice may be much wider if the community is large and cosmopolitan like the Catholic Church, or one of the larger caste communities of India. The first sort of community, the small or localized one, might thus represent a biological subpopulation, and there is some evidence to this effect. We might also belong to an occupational community, and, in fact, caste divisions often include occupational divisions. In Western countries, membership in a profession such as medicine or law or teaching has undoubtedly influenced the marriage pattern although the biological effect is probably slight, in spite of the much publicized endogamy of Hollywood. Economic and social status may have had their effects in some societies. European royalty was, in a sense, a caste community for a time and some biological effects have been attributed to this and to the barriers to intermarriage between nobles and commoners, or between rich and poor. But such communities are generally ill-defined and subject to dislocation by conquest, technological change, and other eddies which keep the gene pool stirred up. Communities of this sort would become subunits of a biological population only if the factors which restrict the choice of mates operate in the same way over long periods.

The practical problem for the student of evolution is how to identify a population in this sense and how to measure its size. Ecologists have developed methods for recognizing and measuring populations of animals and plants; and anthropologists, demographers, and others interested in the natural history of man (who are essentially human ecologists) have applied these with many inventions of their own to the study of human populations. All have this feature in common: they must recognize population units partially isolated from others, and the isolating factors have usually been spatial or geographic ones. Thus, the definitions of territories occupied, habitats, tribal boundaries, and the like have always had an important place in the identification of subpopulations of a species. The development of adaptations to local conditions leads to genetic diversification and thus to the formation of geographic races to which the races of man correspond.

The application of genetics to such studies required an addition and refinement in our ideas of the essential nature of factors which partially isolate populations from each other. To have evolutionary effects, isolating factors must impede the flow of genes between populations by reducing the probability of intermating. *Isolation* is the condition for evolutionary separation, but again it is relative isolation, the steepening of the gradient by which a probability diminishes. The efficiency of isolating factors may thus be evaluated by comparing the frequencies of the same genes in adjacent populations, provided that the genes have the same probability of survival in the two situations, that is, that their adaptive value is similar in the populations being compared. This would exclude, of course, the use of such genes as that for sickle cells from use as criteria of reproductive isolation between populations with different probabilities of infection by malaria.

In human populations the usual factors leading toward isolation between parts of a species in nature (distance, natural barriers, and so forth) are supplemented by social agencies developed peculiarly by man himself. Language, custom, religion, economic organization, social class—these play increasing roles as societies evolve. In modern societies social isolation tends to replace other isolating factors of the natural environment.

THE JEWISH COMMUNITY OF ROME

Some of the problems and ideas presented by isolated communities may become clearer to the reader if I describe a study of one in which I participated together with my son Dr. S. P. Dunn, an anthropologist. It dealt with the small Jewish community in the old ghetto district of Rome. Now all Jewish communities are interesting not only because of the long documented history of the Jews but also because the Jews, as a people, have retained an identity throughout a most turbulent period of human history. All the cards seem to have been stacked against such a retention. An analysis of the cohesive forces which enabled them, in dispersion, usually as minority groups, to keep and transmit an ancient cultural and religious tradition is a fascinating problem for the sociologist or historian of culture. To the biologist the existence of such a group poses questions fully as interesting. If there is a cultural or social unity in Jewry as a whole or in any single community of Jews, has this had biological effects? Do the rules requiring Jews to marry Jews provide effective isolation from the surrounding non-Jewish population? If they do and if this can be tested by comparing the gene frequencies found within the Jewish community with those of the surrounding population, then we have ways of evaluating some of the biological effects of the social and cultural developments which have been so prominent a feature of the last stages of human evolution, and which are likely to influence our future.

Until recently, in Europe, the typical social unit of the Jews was the community, usually small. In northern and eastern Europe this became known in Yiddish as the *Shtetl,* the "little town." In Italy it was the *università,* which has now become the *communità,* a secular organization recognized by the state as representing the Jewish inhabitants of a town or region. These from the days of the Roman Empire were given status as a group, enjoying certain privileges and subject to certain restrictions. With the advent of Christianity, this view of the Jews as a people apart culminated in the enclosure of many such communities in a walled ghetto. In Italy the oldest of these communities was in Rome, where Jews from Palestine had been settled since about 160 B.C. The Roman colony received its largest accession from prisoners of war after Titus captured Jerusalem in A.D. 70. At the dissolution of the Empire and during the Dark Ages small groups of Jews were settled in villages in central and southern Italy, but with the establishment of ghettos in the sixteenth century, all were required to withdraw to one of these enclosures. The ghetto at Rome, a small walled area of some six city blocks inhabited by several thousand Jews, was closed by Pope Paul IV in 1554 and kept closed for 300 years. The intent was clearly both social and reproductive isolation, for there was prohibition of intermarriage and a curfew law, and employment of Jews in Christian homes was forbidden. After the opening of the Roman ghetto by Garibaldi's troops in 1870, this area between the foot of the Palatine and the Tiber, together with the adjacent region across the Tiber (Trastevere), the probable area of first Jewish settlement, remained the social and religious center of Roman Jewry.

When we arrived in Rome to study this community, the first question to be faced was whether we could identify a social community of descendants of the ancient one which had been enclosed in the ghetto. We found a group of families, with names recognized by Romans as "ghetto names," inscribed as members of the "comunità ebraica di Roma," with residence pattern centering in the ghetto area, using the social facilities (temple, Jewish school, health service, Jewish hospital, orphanage, and old people's home) which also were in the same ancient area, often in the very buildings of medieval or earlier date which for many centuries had been inhabited by Jews. The marriage and birth records showed that these families were prevailingly of Roman Jewish origin, marrying chiefly among themselves. So few had married outside the community (and these could be identified) that we decided that we had found what we were looking for, an endogamous community, with a high degree of reproductive isolation, in the midst of a large modern city. As near as we could estimate, this population was rather small—about 4000 people. We called it the "nuclear" community to distinguish it for our purposes from the total Jewish community of Rome, which had about 12,000 inscribed members. The remainder of the larger community were largely Jews of non-Roman origin.

We set about getting estimates of gene frequencies from members of the nuclear community, using our genealogical records as criteria for membership. This meant that we had to persuade by various means as many members as possible to present themselves for examination by families, for it was our purpose to diagnose genotypes by using only genes which could be securely identified by testing blood, saliva, and urine. It is more effective to make tests on parents and children at the same time, since the phenotypes of the children ofter reveal the genotypes of the parents and vice versa, and one serves as a check on the other and on the accuracy of the methods. We succeeded in testing about 700 people, that is, about one-fifth of the membership of the nuclear community. This was enough to permit some judgment on one of our biological questions: whether frequencies of some genes differed between the nuclear community and the surrounding population. For example, the proportion of people of blood-group B was more than twice as great within this community as in the Italian Catholic population generally. In our community the proportion was 27 percent; nowhere else among Italians tested did it exceed values of 10 or 11 percent. There cannot have been very much exchange of genes through intermarriage between Jews of this community and Italian Catholics; if this had happened the frequencies of B would tend toward similarity in the two populations.

Some other Jewish populations examined have commonly shown higher frequencies of B than the populations within which they live. It may be an indication that the ancestral population from which the Jews derive was one with a high B frequency. This would not be an unreasonable assumption for an eastern Mediterranean population. But it would also suggest retention by separated Jewish populations of an ancestral genotype. Perhaps two millenniums of dispersion among other populations have not had the effects that would commonly be expected, namely, intermarriage and amalgamation. Perhaps 70 generations or so is too short a time to permit the effects to show; but, perhaps, and this seems more probable, life in communities isolated by prohibition of outmarriage does, even in modern times, constitute fairly effective reproductive isolation. Certainly it appears to have had this effect in Rome.

The proportions of persons of blood-group B in the Roman community is high even when compared with other Jewish communities. This may reflect an effect of size, since small communities often exhibit peculiar extremes of gene frequencies, usually attributed to chance fluctuations (drift) by which a small eddy of the gene pool gets separated from the main body and retains its accidentally acquired peculiarity. This seems more reasonable in the present case than to suppose that persons of group B were favored by selection, although, of course, selection cannot be excluded.

In the case of another peculiarity of this community, drift seems altogether a more likely explanation. One of the Rh alleles (r′, a form of Rh-negative), which is usually rare in Europe, attains a frequency of over 5 percent in this community, higher than in any Jewish population known (although in not many has this allele been looked for) and much higher than in Italy generally. Gene flow by intermarriage with a population with a much higher frequency of this allele cannot account for it since such a population is not known.

Whatever the causes of these differences in gene frequency, they are at least evidence that in this case social isolation has been biologically effective. Effects of this sort have been observed frequently enough among both animals and men to have led to the use of a special term for them. A small population subject to geographic or social isolation from the surrounding population is called an *isolate*. In using the term, one should remember that isolation, in human populations, is always relative, expressing a difference in the probabilities of mating within, as opposed to outside, the group. Isolating factors in human populations may be geographical or social. Differences in language, religion, or custom may act as social impediments to mating. In some societies race prejudice—a social acquirement—may be an isolating factor, as in the southern United States. The Roman Jewish community has been a social isolate formed under the influence of a variety of historical factors both religious and economic.

The Jewish population of the world has of course been living in dispersed communities for some 2000 years. Members of many of these communities have recently settled in Israel. Advantage has been taken of this opportunity by Israeli and other scientists to compare the gene frequencies and physical traits of members of these resettled communities before they become merged in the larger national community. Some of these observations are reviewed in a recent book, edited by E. Goldschmidt, reporting the results of a conference held in 1961 at the Hebrew University, Jerusalem. The frequencies of the ABO blood groups show some interesting differences between these communities. Thus, the "Black Jews" from Cochin, India, who suppose themselves to be descended from refugees who escaped from Palestine after the fall of the Second Temple (A.D. 70), show closer resemblance to their Hindu neighbors than to other Jewish groups. Ashkenazic Jews from eastern Europe and Sephardic Jews from southern Europe and North Africa do not differ very much from each other nor from Oriental Jews. The Yemenite Jews, with a high frequency of blood-group O and low frequencies of A and B, differ in this respect from the other communities but resemble the Yemenite Arabs. In general, long-separated Jewish communities vary among themselves but often show some resemblances to the populations among whom they live. The ABO blood-group frequencies, as noted earlier, often vary widely between populations known from other evidence to be racially similar, and judgments about racial homogeneity or diversity of Jewish groups should probably not be based on these genes. The Rh gene frequencies are somewhat less variable and show two interesting features; the east European Jews (Ashkenazim) have less of the Rh-negative gene and more of the R^o (the "African" gene) than the populations among which they have lived, just as they also have more blood-group B. Jewish communities are, thus, probably descended from early Jewish emigrants from the eastern Mediterranean who retain in their genes evidences of that origin, the degree of retention depending on the degree to which reproductive isolation from the host populations has been effective. In some communities, like those of Cochin or Yemen, either this isolation broke down through intermarriage or else some of the native population were converted to Judaism, which would have had a similar effect. In other communities, such as those of eastern Europe, this took place to a lesser extent.

One interesting comparison is between the Roman Jewish community and the Jews from Tripolitania now settled in Israel. The ABO, MN, and Rh gene frequencies, the only ones available from the Tripoli community, are almost identical with those in Rome. One would guess that they had recently been drawn from the same population. In fact, the community at Tripoli received in the seventeenth century emigrants from Livorno (Leghorn), Italy, probably similar in derivation to the Jews in the small community in Rome. This emphasizes the fact that when comparisons are made of gene frequencies between recently separated communities one is on safer ground in inferring from similarities an historical connection. Both communities retained, in different host populations, the high B frequency characteristic of the Roman Jews.

The comparisons between Jews of different communities made by Professor Sachs utilized variations in finger print patterns. These are inherited, but the individual genes have not been identified, so comparisons of this sort are more like those of the physical anthropologists based on phenotypes. The comparisons, based on large numbers of observations, show a surprising degree of resemblance between members of separated Jewish communities, and differences between these and the host populations.

On the other hand, the G6PD deficiency is extremely rare (0.04 percent) in Ashkenazim (Jews from Northern and Eastern Europe), whereas in Kurdish Jews from the Iraqui border it has the highest frequency (70 percent) ever found in any population. Similarly, familial Mediterranean fever, a rare disease due to a recessive gene, is virtually limited in its occurrence to Sephardic and Mediterranean Jews. It is too soon to evaluate the roles played by selection in different environments, genetic drift, and outmarriage to different populations in causing these large differences. Both of the above-mentioned genes are subject to strong selective pressures which might cause more rapid changes than the blood-group genes appear to show.

We cannot yet assess the degree of biological unity retained in dispersed Jewish communities. It can be said now that it is, for certain genes and phenotypes, greater than would have been expected from the long history of movements of Jewish groups through other populations —evidence, perhaps, of adherence on the part of those who remained members of Jewish communities to the practice of marriage within the faith.

OTHER RELIGIOUS ISOLATES

Two other studies of what we may now call "religious isolates" are of special interest. Both bear testimony to the force of religious rules about marriage.

In one of these investigations, Professor Bentley Glass of Johns Hopkins University, with other colleagues, tested most of the members of a small community of Dunkers, descended from German Baptist Brethren who emigrated to the United States beginning in 1719. In 1881 the sect split into three groups, the smallest and most retentive of the original practices being the Old German Baptist Brethren, numbering now some 3500 members. In one community of this sect in Franklin County, Pennsylvania, numbering about 300 people, three blood-group gene systems (ABO, MN, Rh) and three other genes affecting external traits were studied in 265 members. When the frequencies of these genes were compared with those characteristic of the general American population within which the Dunkers live, and of the West German population from which the Dunkers had come, striking differences were found. Blood-group gene *B* had practically disappeared from the community, gene *A* was much more frequent than in either West Germany or the United States, and gene *O* less frequent. Over 44 percent of the Dunker community were of blood-group M as compared with 29 percent in both the United States and West Germany. The frequency of three of the genes diagnosed from physical traits of ears and hands also differed between the isolate and the parent and host populations. Glass points out that the differences could not have been due to inflow of genes from other populations since the frequencies in the isolate are outside the range of those of populations with which the Dunkers could have come in contact. He attributes the differences to random drift due to chance fluctuations in the small sample of genes in this group, the differences being maintained by marriage chiefly within the group. For example, only one person whose parents were born in the community is of blood-group AB. If the parents had left the community a generation ago, the gene *B* would have disappeared entirely. Four other carriers of gene *B* from another community subsequently married into the isolate, but the existence of gene *B* in the community is obviously precarious. When the community is small and chiefly endogamous the frequencies of individual genes may be subject to such accidental fluctuations, some tending to get fixed and others to get lost. This is purely an effect of size and could happen without selective advantage or disadvantage of the genes themselves; but, as pointed out before, such accidents constitute opportunities which result in evolutionary change only if permitted by circumstances which persist for some generations.

Our second example of isolation by religious and social rules is interesting for another reason. A Hindu caste in India is either a single endogamous (inmarrying) group or a collection of several endogamous groups. The Brahman caste, for example, consists of some million members, divided into four geographical groups, each of which is further divided into subdivisions, and these in turn into endogamous communities. Each member is permitted to marry only within his own community, but in some communities he may not select a mate from within a certain narrow section of his close relatives. In some Brahman communities, for example, a boy may not marry the daughter of his mother's brother, his cross-cousin; in others cross-cousin marriage is strongly encouraged. Such systems lead to inbreeding, but the existence of exogamous sections moderates the degree of inbreeding. The rules would, if rigidly followed, produce complete isolation between the different castes, even though their members live in close association with each other, and would tend to isolate different communities within the caste. This system has been operating for more than 1000 years and some of the communities have been isolated by social rules for 100 generations or more. Some of the castes are very large, numbered in millions; others are small, although none which has been studied is very small, less than about 30,000.

One interesting evolutionary question is clearly this: what biological effect has this system had on a population divided into isolates of the above sort? Does it, for example, tend toward the formation of different races within a single area? Such a tendency should first become detectable by differences in the frequencies of the same genes betwen different endogamous groups.

Dr. L. V. Sanghvi of the Human Variation Unit of the Indian Cancer Research Center of Bombay, and Professor V. R. Khanolkar, director of the Center, undertook to study this question by comparing the frequency of several genetic characters, including some blood antigens, in five endogamous communities in Bombay. Striking differences in some of the genetic characters between different endogamous communities were revealed, some of them as large as those found for certain gene differences between American whites and American Negroes. Subsequently Dr. G. M. Kurulkar of Bombay measured the same subjects, and the morphological characters also revealed differences between the endogamous groups. When the five groups were compared first by genetical differences and then independently by morphological differences, the two resulting arrangements were similar. Thus, whether judged by genes or by the kinds of phenotypes recognized by physical anthropologists, some of the caste communities appear to differ as races differ. One cannot conclude directly from this that the long continued application of the marriage rules of the castes have actually been *race-forming* influences; they may have been the agencies by which ancient racial differences between the castes or communities have been preserved. But the present state of the differences leaves no doubt that the marriage rules have been effective as factors contributing to reproductive isolation. Such social forces are, at the least, potential race-forming factors, by which a variety of group genotypes is provided upon which other evolutionary forces may act.

NEGRO COMMUNITIES IN THE NEW WORLD

One of the great ethnic dislocations of recent times was the forced migration of Negroes from Africa to the Western Hemisphere. In some places they were largely absorbed into the native population by miscegenation or intermarriage, as in Mexico; in others this process took place to lesser degrees. In the United States, we have relatively little precise information on the degree to which European genes have diffused, by interbreeding, into the Negro population. The best estimates, by Glass, indicate that the Negroes of Baltimore and New York City have admixtures of European genes amounting to between 20 and 30 percent. Obviously, Europeans and Negroes have not been reproductively isolated. The rate of flow of European genes into the Negro population is estimated at around 3 percent per generation during the dozen or so generations of contact in America. No really good evidence exists for the popular supposition that the American Negroes of today have inherited some of their genes from American Indian ancestors. The amount of gene flow, if it has occurred, is too small to be measured by present methods.

Here and there in the New World smaller localized populations of Negroes have retained, in communities, assortments of genes resembling those which had come from Africa in their ancestors. Some of these communities are found on the strip of lowland plain extending some 50 miles inland along the Atlantic coast of South Carolina and Georgia. This is the home of the Gullah Negroes, so called because of the dialect spoken by them, which retains many African features. Physically, too, they appear more "African" in skin color and facial features than the average Negroes in other parts of the country. Estimates of the gene frequencies of a sample of Gullah Negroes from Charleston, South Carolina, as studied by Dr. W. S. Pollitzer of the University of

North Carolina, do, in fact, show them to be much closer to present-day west Africans than are the Negroes of New York or Baltimore. The flow of European genes into this population is too small to be detected. They have, however, diverged somewhat from the present-day west Africans, but how much of this is due to gene flow, to genetic drift, or to selection is unknown. One of their African peculiarities is the retention of a higher proportion of sickle-cell genes than other American Negro populations in which this trait has been studied. Some part of this may be due to their long residence in the malarial area along the coast.

A small Negro community on one of the so-called Sea Islands (James Island) of this region was studied by methods similar to those described for the Roman community. It consists of about 150 families, interrelated, probably with considerable consanguinity. Indications are that it has for some time been reproductively isolated from the surrounding white population and partially so from more distant Negro populations as well. There is as yet no secure indication within it of genes received from Europeans. The frequency of such typically African genes as R^0 and of sickling is high, and, in addition, the genes for thalassemia and for the abnormal hemoglobin C have been found in some of the families, together with two variants of the MN blood-group system, known as the Henshaw and Hunter antigens, which are rare even in African populations and not found in Europeans. Although the study of this community has been begun too recently to permit any final conclusions to be drawn, it appears that here we may have a little pocket in which African genes have been retained and concentrated by reproductive isolation.

Dr. Lester Firschein, now of New York University, found that a somewhat similar situation seems to exist in the so-called Black Caribs along the gulf coasts and islands off British Honduras. Here the contrasts between cultural and biological inheritance are striking, for, although the genes of this community are prevailingly African, their language and many of their customs and ways of life are Indian. The Black Caribs originated in African slaves deported to this coast in the eighteenth century after a few generations of residence on a West Indian island (St. Vincent). The factors leading toward reproductive isolation of this population are partly geographic (they are coastal peoples), possibly partly adaptive, since, with high frequencies of the sickle-cell gene (absent from American Indians) and perhaps other genes conferring higher resistance to malaria, they were able to succeed in a region which was less hospitable to other populations. An important observation of Firschein was that those who carry the sickle-cell allele had a higher net fertility than those with the non-sickle genotype. The selective advantage of the sickle-cell allele as expressed in the malarial environment was in part responsible for its retention in high frequency. Here, too, the relatively small size of the isolated population has led to endogamy, with preservation of an ancestral gene pool in the midst of a culture pattern acquired in the New World.

EFFECTS OF VARIATION IN THE SIZES OF POPULATIONS

Where a total population, like that of man, is broken up into intermarrying groups which are reproductively isolated from each other, even partially, the size of each subpopulation or isolate must be limited. The pool from which genes are drawn must be smaller than that of the whole species. Several effects might be expected to follow from this. First, when the circle of possible mates is restricted, as in an isolate, random mating should result in more marriages between cousins than in a larger open population. This is purely an effect of numbers: if you have to find your mate among a small number of persons of similar descent you are more likely to marry a relative, quite apart from any choice on your part. And, where you marry a relative, the chance that you and she (or you and he) will have similar genes is greater than if you are not related, simply because some of the genes in each of you came from an ancestor common to both of you. If you and your wife are first cousins, that means you had a pair of grandparents in common: instead of the two of you having each four different grandparents, eight altogether, you have only six different grandparents between you. Such marriages are *consanguineous* since they involve common blood. Blood as used here, of course, is just an old-fashioned expression for genes, since you inherit not blood but genes. A continuous system of consanguineous marriages constitutes inbreeding, the degree of which can be measured by the reduction in the number of different ancestors possible if no inbreeding took place. Or it can be measured by the increase in the probability that the two spouses will have genes in common, that is, that the genes of one will show greater correlation with those of the other than in two persons chosen at random.

Now an inbreeding system deliberately or experimentally imposed on populations of animals and plants is known to produce two main effects on the distribution of genes. Within inbred lines or families descended from common ancestors, individuals with two like alleles in any pair (homozygotes) become more frequent than under outbreeding, and this follows automatically from the mechanism of gene segregation. Mendel himself first showed this to be a true inference from his law of segregation. Under a continuous system of the closest possible inbreeding, namely, when male and female gametes from the same individual unite in self-fertilization (possible in many plants which are true hermaphrodites), the proportion of heterozygotes for any pair of alleles is halved in every generation of descendants from a heterozygote. This automatically produces the second marked effect of inbreeding, namely, the differentiation, in the alleles they contain, between *different* families or lines. A heterozygote *Aa* produces by self-fertilization *AA* and *Aa* and *aa* individuals. If, thereafter, *AA* can produce only *AA* individuals and *aa* only *aa* ones while *Aa* produces always *AA, Aa,* and *aa,* the homozygous classes will increase at the expense of the heterozygotes (which continually break down into homozygotes), After a time under such a system we shall have mainly *AA* and *aa,* and the proportion of *Aa* will approach zero. Lesser degrees of inbreeding, such as between brother by sister, or between first cousins, produce a similar effect but more slowly in terms of generations.

Consequently, in a small isolated population we should look for: 1) an increase in the proportion of cousin marriages over that in the larger surrounding population; 2) an increase in the proportions of homozygotes. The first effect was undoubtedly present in the small Roman community; cousin marriages between persons of the same surname were frequent, but we were unable to get a precise figure for it because of inability to get complete records of the grandparents of the persons we studied. Establishment of cousinship is a three generation affair and, in a community such as this, in which no special importance was attached to cousin marriage, statements concerning it would have to be checked for grade of relationship (first, second, third, or half cousin, for example) from records which we could not get. Studies of other small communities, especially of Catholic ones in which cousin marriages can be performed only with a written dispensation which becomes a part of the marriage record, leave no doubt that the frequency of cousin marriages is higher in small communities. The importance of a precise figure is that the relation between the consanguinity rate and the size of the group within which marriages are contracted at random is regular enough to permit a rough estimate of the size of the isolate from the cousin marriage rate. Any such relation would break down, of course, if the marriage pattern is controlled by social rules specifying the relationship between the parties to a marriage. Thus, in endogamous Hindu communities in India studied by Sanghvi and others, marriages between cousins make up from 6 to 11 percent of all marriages, a high rate of consanguinity even though some of the communities are large.

Even these high rates would have insignificant effects on the proportion of homozygotes for *common* genes such as those for the blood antigens and others used in comparing populations. Sanghvi failed to detect any such effect in the Indian communities; we failed to do so in the Roman community; nor is it evident in the small Dunker community. In fact, there is little likelihood that increases in the frequency of homozygotes for common genes will be detected in human communities.

With rare genes it is a different story. These have low probabilities of getting into homozygous combination under random mating, but this probability is greatly increased when the descendants of the same heterozygote, such as cousins, marry. A good illustration of this is the recessive gene for juvenile amaurotic idiocy, which in Sweden has a frequency of 0.005. This means that in the population as a whole about 1 percent of the people are heterozygous for this gene (Aa). The probability of mating between two heterozygotes $(Aa \times Aa)$ would then be 0.01×0.01 or 0.0001 and from such matings the probability of a homozygote *aa* would be ¼ or 0.25, so the appearance of such homozygotes by chance would be 0.0001×0.25 or 0.000025, that is, about 1 per 40,000, which is the observed frequency of this disease in Sweden. But suppose now that cousins marry who had a common grandparent Aa. The chances of descent of the gene *a* in such a pedigree would be as in Fig. 11. The chance of birth of a child homozygous for this gene (*aa*) is about $\frac{1}{64}$. If *a* has a frequency of .005, as in Sweden, then the risk of *aa* from cousin marriage is about $\frac{1}{3200}$. This is to be compared with the chance of birth of $\frac{1}{40,000}$ from matings contracted at random. Restriction of marriages to those within a small isolate, in which the gene occurs at all and in which cousin marriages are common, greatly increases the chance that it will appear in homozygous form. If the gene is common,

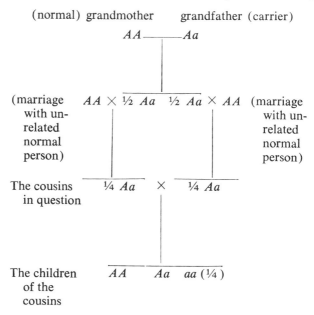

The chance of birth of a homozygous *aa* child from this mating is computed as follows: chance that any cousin should be $Aa = \frac{1}{4}$; chance that two such cousins should marry $= \frac{1}{4} \times \frac{1}{4} = \frac{1}{16}$; chance that their child should be $aa = \frac{1}{16} \times \frac{1}{4} = \frac{1}{64}$ or 0.015625.

Fig. 11. The chance of birth of a child with juvenile amaurotic idiocy from a cousin marriage between grandchildren of a person heterozygous for the gene.

say, with a frequency of .25, repeating the above computation and *not* neglecting the chance that normal unrelated persons may carry the recessive gene yields a probability of its becoming homozygous in the offspring of cousins of about 0.07, as compared with the random expectations of $0.0625 = (0.25)^2$. This is too slight a difference to be detected, remembering that only a portion of marriages in the isolate will be between cousins.

The result of all this is that the detection of the effect of community size or of the inbreeding that accompanies it upon the frequency of homozygotes is a difficult matter and no really good evidence on it exists for human communities. One often hears it said "Oh, that's an inbred community in which the people are as alike as two peas in a pod and it's full of degenerates." Such statements usually do not stand up after investigation. The people of the small Roman community, for example, were variable and not to be distinguished by any obvious abnormalities from the rest of the natives of Rome. Other investigations of small communities have led to similar opinions. Small minority groups are likely to be marked by superficial peculiarities of dress or habits or economic status, just because they are minorities maintaining themselves against the influence of the dominant culture that surrounds them, but these are not to be confused or mistaken for biological effects resulting from isolation or inbreeding. Only biological observations could establish such effects, and there are not yet enough of them to permit conclusions to be drawn.

RANDOM GENETIC DRIFT

Another effect of small effective size (small number of reproducing individuals) in a population is to make it liable to accidental or chance fluctuations in the frequencies of its genes. When these are accumulated over a number of generations into significant departures from the original equilibrium, the evolutionary force which subjects the population to change is known as random genetic drift. The theory and its application to evolutionary processes was developed chiefly by Professor Sewall Wright, who laid much of the groundwork of theory in population genetics. Drift, thus, sometimes is called the "Sewall Wright effect," or after John Gulick who first suggested the idea, "the Gulick effect."

Examples of changes in gene frequencies in human populations which may be due to drift have already been given. I say "may be" because strict proof of this effect is difficult, perhaps impossible, to get in human populations in which the other forces, such as selection, which play upon them are difficult to evaluate separately. Wright's idea was that a large population, subdivided into smaller local breeding groups, offers the best opportunities for natural selection to choose among the groups those which best meet the environmental requirements. Differences among the groups depend both upon their reproductive isolation and upon their effective sizes, the smaller being more likely to vary. The final decisive force is intergroup selection.

In spite of the difficulties of testing the applicability of this view to man, it is an attractive idea to anthropologists and other students of human evolution. For, in the distant past, our ancestors probably lived under these conditions in small isolated or partly isolated bands. Most human beings were and many still are members of breeding communities limited in size by a variety of factors both natural and social.

One of the most interesting pictures of the structures of primitive populations and the forces that play upon them to produce intergroup variety and change has emerged from recent observations of Australian aborigines. Here the whole continent, at the time of first European contact, was inhabited by a population still at a Stone Age level of culture, living only by hunting and gathering. It is estimated that the total population at that time consisted of from 250,000 to 300,000 persons living in some 575 scattered tribal units relatively isolated reproductively from each other. A tribe held together by internal cohesion and attachment to a hunting and gathering area thus had an average size of about 500 persons. Its effective breeding size was, of course, much smaller.

Now, thanks to the work of the American anthropologist J. B. Birdsell and his Australian collaborators R. T. Simmons, J. J. Graydon, N. B. Tindale, and others, we know something about the structures of these populations and the frequency of some of the genes in them. From them we have suggestions concerning some of the forces responsible for the variations in gene frequency. Genetic drift is one of these. I am grateful to Professor Birdsell, Mr. Simmons, and Dr. Graydon for permission to quote from the unpublished results of their extended bloodtyping work in western Australia in 1952–53. They succeeded in testing 1850 full-blooded aborigines in 25 tribal isolates varying in size from 20 to 300 individuals. When the frequencies of the genes of the ABO, MN, and Rh blood-group systems are arranged by tribes two general results of great interest emerge. One is a pattern of frequencies changing with distance, which I have referred to earlier as clines. The other, more pertinent in the present context, is the great and irregular variety in gene frequencies among the isolates, sometimes departing widely from the value expected on the clinal gradient. For example, blood-group gene *O,* although increasing from 47 percent in the south to 87 percent in the north, has a maximum difference between adjacent isolates of 30 percent. The gene *m* varies in different isolates from 2 to 37 percent, with differences between adjacent isolates amounting in several instances to as much as 15 percent.

The Rh genes are not so well known but at least four

of the Rh-positive phenotypes, R^1, R^2, R^0, and R^z (the Rh-negative gene r is absent), show similar abrupt differences between adjacent isolates.

Professor Birdsell writes (*in literis*): "At all three loci, the frequencies seem under broad clinal control and presumably hence are responsive to adaptive differences. Even so, in each instance, some isolates deviate widely and significantly from their expected position in the clinal topography, and suggest that such abrupt fluctuations may locally represent isolates in which drift has occurred. The data do not suggest one or the other process working exclusively, but rather that in the Australian isolate the drift may occasionally produce regionally localized fluctuations from expected values. These do not seem due to sampling error in all cases and suggest that drift may sometimes be a minor factor in the evolution of populations of this structure."

Certainly the small sizes of the tribal breeding units would render them sensitive to random fluctuations, and a variety of group genotypes would be offered for other evolutionary forces to act upon. We should remember in this connection that the whole of Australia was an anciently isolated region. The so-called "Wallace's line," the strait between Bali and Lombok north of Australia, was first detected by Alfred Russell Wallace in the abruptness of change in the animal and plant forms from the Malayan area in the north to the Australasian in the south. But, once isolated below this geographic barrier, the fauna of Australia, including its human inhabitants, although it became peculiarly different from its ancient relatives above the line, did not become uniform. The type example of microevolution in man might well be taken as this great variability in the genetic constitution of the Australian tribes.

Professor Glass has expressed very well the possible function of genetic drift in evolutionary changes at this level. "In this way, then, we may picture the creative role of random genetic drift. Like the randomness of mutation which produces a hundred or a thousand deleterious genes to one that aids the organism in its struggle for survival, genetic drift may establish a hundred or a thousand deleterious mutations, fated eventually to be eliminated, before it aids materially in the initial establishment of a mutation of value. And, like recombination—indeed, through recombination—genetic drift provides those unique genotypes that characterize small closed populations, and upon which selection can act with so much richer a variety of outcome than when it is limited in material to the phenotypic uniformity of the large, panmictic population. In Sewall Wright's own picturesque imagery, the way from one adaptive peak to another leads through the valley. The shadow of selection broods darkly over the evolutionary way, and only the strong or the fortunate escape, only those forearmed against the exigencies they meet. Stabilizing selection adapts more perfectly only to those conditions that have existed up to now. It may be that solely among the less highly adapted, small isolated populations there will occur the genotypic system that fits the new conditions. In helping to create a diversity of such systems, random genetic drift plays its part."

VII

A Look Ahead

The intellectual revolution of the nineteenth century, which was initiated by Darwin's book of 1859, has by now pretty generally affected our thinking. We have become accustomed to thinking of living things as part of one continuum, the history of each species extending back in time to the origin of life itself. Nor do we object seriously to regarding ourselves as a part, albeit the culminating part, of that long historical process. Most people, however, accept it as history, something that is over and done with. Since we do not know in detail the actual steps, with dates, or the guiding forces (except in outline) which brought about the emergence of man as a species, we tend, on the whole, to accept the view that man, as man, appeared on the earth 600,000 or 700,000 years ago—and that was that.

That, however, would represent a very incomplete appreciation of evolution. Living things do not stand still at any time. The meaning of evolution is change, the continuous process of adjustment arising out of continually re-created genetic variety which is sifted by the variety of natural conditions on this planet. Evolution is not something accomplished, but something which is continually occurring, a living, dynamic process.

It is so no less for our species than for others of shorter lifespan in which the flux is easily observed, as the populations of bacteria or viruses which change almost from day to day, or the garden plants which appear in new forms annually in the seed catalogues.

Even though no major changes in the structure of

man have been noted in the more than half-million years of his existence (in the sense that he has shown no tendency to split into different species), there is ample evidence that human evolution is occurring today. We see this in the disappearance of some racial groups and the fusion of others to form new complexes, like the American Negro, the "Cape Colored" of South Africa, the modern Hawaiians, and others. We see old groupings, like the Jews, separating into populations showing new forms of diversity; segments of populations isolated from the main body developing peculiarities of their own or merging with the host population and thereby changing it. We see gene frequencies changing by man's own action, as in the case of sickle-cell anemia and thalassemia. Most important, we have good evidence that human genes are changing by mutation today; and mutation supplies the raw material for evolution.

The changes in the races or natural varieties of man have been called microevolution. Those of us who spend our time studying them believe them to be, on however small a scale, evolution itself. We have no doubt that these kaleidoscopic processes of change will continue. There seems to us to be no prospect at all that the human species will be less various than it is now. Evolution, it seems to us, will stop only with life itself.

What then of the future? How shall we, the human population of this globe, change? How would we change if we could gain some control of our biological composition as a species?

HUMAN IMPROVEMENT

The Social Control of Human Evolution was the apt title of a book by W. S. Kellicott published in 1911. Today the phrase sounds a little pretentious; and we realize now that the confident hope expressed in it was based only on the first fruits of the new knowledge which came from the dual stimuli of Darwin's and of

Mendel's work. Nevertheless, it voices an ancient hope. It might well have been used for that section in the fifth book of Plato's *Republic* in which Socrates asks Glaucon how he would breed better men. Would he not, suggests Socrates with his usual leading question, mate best with best, as he would do with his sheep or horses? This was the essential intent of Sir Frances Galton, Darwin's cousin, who began asking such questions in the 1880's and eventually elaborated them into the proposal which he called eugenics. The general program of eugenics, as set forth by Galton and his many followers, involves some form of social control of human evolution by selection, that is, by influencing the probability that particular genes will be represented in the offspring. It always involves a judgment about which genes are desirable and which undesirable. The undesirability of perpetuating certain genes, such as those for crippling diseases, idiocies, and severe mental or phyical defects, would be generally agreed to. We should prevent the transmission of those genes if we can. As Mr. Justice Holmes remarked in giving judgment on a case involving compulsory sterilization of an idiot, "Three generations of idiots are enough." The proposals for this sort of control, usually referred to as negative eugenics, sound simple enough. The difficulties in application arise from the fact that in order to stop the transmission of a gene one must first have identified the undesirable phenotype, such as a mental defect, as due to a gene and not to some other cause. This takes research which has been accomplished only for some defects, most of them those which are already kept rare by natural selection.

When the gene has been identified, if it turns out to be both rare and recessive (and most new mutations are recessive), another difficulty is encountered in getting rid of it. If we can recognize it only in homozygotes and aim our negative eugenics program at preventing them from passing it on, we shall miss the great majority of such genes in the population, for these are in the heterozygotes, the "normal" people. A little arithmetic will make this clear. Suppose the rare recessive gene we want to eliminate is that for juvenile amaurotic idiocy. The frequency of this allele is estimated to be 5 per 1000 in Sweden, the source of our best evidence. This means that in the Swedish population of 7,500,000 people, each with 2 alleles at this locus, there are 75,000 such defective alleles. The equilibrium formula (Hardy-Weinberg) tells us to expect that about 375 of these are in homozygotes and 74,625 in heterozygotes. To eliminate all homozygotes would get rid of only 0.5 percent of the bad alleles. In the case of this disease, nature already prevents the transmission of genes from the homozygotes since all die before the reproductive age. To have any effect, our program would have to be aimed at reducing the transmission of these genes by the heterozygotes. That means we should have to learn how to detect the normal carriers *before* they passed on their genes.

Some other bad recessive genes are not so rare as this, but it is always true that the majority of recessive genes, except those which are very common indeed, are in the carriers, the heterozygotes. Eliminating homozygotes becomes less and less effective as the gene in question becomes less common.

There is no such difficulty with genes which are clearly dominant or those which can be detected in heterozygotes by indirect means. Sickle-cell anemia, as we saw earlier, could be quickly reduced in frequency if all heterozygotes refused to have children or were prevented from doing so. Negative eugenics could be made to work with adequate knowledge and the social sanctions which would obviously be required.

Eugenic proposals of a positive sort—the sort of thing that Socrates was suggesting—are not so clean-cut. Encouraging the mating of "best by best" is easier advised than done. The ideals of what is "best' in man have

undergone many changes since Plato's time and differ
from society to society. Although we might get some
agreement that virtue, Socrates' ideal, was a desirable
goal, we should find that, since it lies in the moral rather
than in the biological domain, we should be unable to
prescribe a program for breeding for it as we breed for
useful qualities in domestic animals.

The judgment of *biological* excellence is less subject
to changing fashions; but it is rendered in the court of
nature which judges fitness or adaptedness. For man,
adaptedness comes more and more to depend upon the
use of his wits to make his environment fit for him. We
can, by use of our inventiveness, live practically any-
where. We should, if we want to supplement natural
selection, learn how to select for inborn mental ability.
The truth is, we don't know how to do it now. Galton
thought we could simply make it easier for bright people
to have more children; but we don't yet know the gene
combinations which make for brightness. For Galton,
success itself was the criterion; social, financial, or intellectual eminence was the proof that the genes were
there. But we haven't yet succeeded in disentangling the
effects of genes on intellectual performance, on emo-
tional stability or the lack of it, from the complex
environmental matrix in which they operate, which may
be different for each human being. Many schemes have
been proposed and some have been tried for increasing
the birth rate of the gifted and the successful: family
allowances, bonuses, selective taxation—but all encounter the instabilities of a competitive economic system based upon changing technology and the shifts of
judgment about what constitutes success. Our efforts to
apply some social control to human evolution fall back
eventually upon reducing the proportion of those men-
tally and physically unfitted for successful adjustment to
our complex technological society. These tend to become a more serious social impediment, not because

(as some have thought, without good evidence) their relative numbers increase, but because the advance of technology itself leaves them farther behind.

If we revert to negative eugenic measures to ease this social burden we should assess the amount of dependence we can place upon it. We should remember that this prescription for alleviating the biological ills of our species was devised at a time when the genetic system, which is subject to deterioration, was not so well understood as it is today. What appeared to Galton as a disease of the evolutionary process, to which we had become subject by the development of societies based on technology, we now recognize as a condition of biological evolution itself. Progress in all species depends on the origination of novelty; mutation is the first step, and the emergence of new combinations of hereditary elements is the second. Novelties always have to be paid for; the evolutionary price is some loss of adaptive fitness in the biologically new that is introduced into an environment to which the old variations had attained some measure of adjustment. It would be asking too much of luck to suppose that internal changes—mutations and their combinations—would be accompanied by just those chance fluctuations in the environment which would provide for the new mutations the conditions required for success. And so the price of variation is maladjustment of some of the variants; but it appears to be a price that surviving species have been able to afford, for all of them must have adjusted their biological systems in this way. Sometime, somewhere, novelties must have paid off, and so this system of trial and error became a part of our evolutionary heritage.

Since we have to live with this system, our ingenuity should be directed to inventing ways of avoiding those of its consequences which hinder to the greatest degree the health and success of our species. We now know that mankind has not only a biological but a social unity arising from the interdependence of human beings in societies and of societies with each other. Consequently, our inventions have to deal with a problem common to all men; they have to be conceived in social terms for social ends. The essential problem is given by the presence in each human individual of single representatives of several recessive genes, each of which, if it were to appear in combination with another gene like itself (for example, in his children or descendants), would have deleterious or fatal effects before birth or in early life. The average, according to recent estimates, may be five or more such genes per person: and it is probable that the average human gene runs a risk of changing by mutation to such a potentially deleterious form of around 1 per 100,000. If there were only 10,000 genes per person susceptible to mutation this would mean that one gamete in 10 might contain a newly arisen deleterious gene, or that, of the half-billion sperms available for a single act of fertilization, one-tenth, or 50 million, of them might contain a new mutant gene of this sort. This prevalence of deleterious genes is what Muller has referred to as "our load of mutations."

At present, our inventiveness is powerless to deal with this problem in the most direct way, namely, by reducing the rate at which new mutations occur. One might ask, of course, whether, if we had this power, we should use it. We might recall that mutation is the source of the variety on which evolutionary progress ultimately depends and we might hesitate for that reason to deprive future generations of this potential source of new genotypes which might be better adapted to new conditions. On the whole, and considering the price to be paid in terms of the loss of fitness accompanying most new mutations, I think we should be inclined to settle for what we have now and to stop the process if we could. We might consider that our species is now heterozygous enough; that somewhere in mankind the genes exist which in proper combinations and in suitable environments provide sufficient genetic opportunities for our descendants. The mutations which have already arisen are no longer restricted to the populations in which they occurred: they are increasingly shared with other populations through migrations and race mixture. Recombination thus arising is also a potent source of variety; and its potentialities for the future evolution of man are far from exhausted.

Speculating about our possible ability to control the mutation rates of human genes is not an idle pastime. We already know several ways in which mutation rates can be increased. Ionizing radiations such as those from x-rays and from atomic fission can do this. We can, if we will, prevent further increases from such sources. And, from experiments with lower organisms, we know that it is possible to counteract the mutagenic effects of some of these agents by treatment with chemicals. It is ironic that it was the threat from devices developed for war that spurred such research. How were we to protect ourselves and our descendants from the effects of radiation that we had literally brought down upon ourselves? First the poison, then the antidote, has been the usual order in technological development, further ingenuity always being required to counteract the effects upon our bodies of previous inventions. The best result of all of this would be the discovery of what actually happens when a gene mutates; with this knowledge we might be in a position to cause specific mutations to occur—or to prevent them from happening.

Our inventiveness would probably lead most immediately to a reduction in the effects of our present load of mutations if it were directed toward detecting such deleterious genes in the persons who carry them. For several harmful genes the scientific problem is already solved. We can identify the persons who will transmit sickle-cell anemia and several similar genes which pro-

duce abnormalities of hemoglobin that can be detected
by electrophoresis; we can identify carriers of thalassemia, of at least three genes which affect the blood-
clotting mechanism, of two which affect the red blood
cells (elliptocytosis and spherocytosis) and of several
with biochemical effects on the composition of blood
and body tissues. The list will undoubtedly be extended
by work now in progress on a number of genes with
effects which have heretofore been regarded as recessive.
The imbecility (phenylketonuria) due to a defect in the
metabolism of phenylalanine is one of these. It has recently been shown that parents known to be heterozygous
because they have had such an imbecilic child can be
distinguished from noncarriers of the gene. They appear
to utilize the amino acid phenylalanine less rapidly than
persons who do not transmit this gene. This substance,
a common component of proteins in foods, can be as-
sayed chemically in the blood serum at intervals after
ingestion of large doses of it. Normal people break it
down rapidly so that the amounts of it decline rather
quickly in the blood; carriers break it down less rapidly;
and homozygotes, who suffer from this form of imbecility, fail to break it down normally and eventually
excrete much of it in the urine. Thus, a sample of blood
and of urine taken from each of the three kinds of persons, when properly analyzed, will reveal to which
category each person belongs. This method can now be
used to detect carriers of the gene who can then be
warned of the risk incurred if they marry similar carriers. It would be applied first where the risk is greatest,
namely, by testing the siblings and relatives of a victim
of this disease. Some other genes for serious mental
deficiencies, the infantile and juvenile forms of amaurotic
idiocy, for example, are likely candidates for detection
in the heterozygous state since the metabolic system
which they affect has been identified although the precise

step in lipid metabolism which goes wrong is not yet known.

Although learning how to detect the carriers of deleterious recessive genes is the first and most essential step in any effort to control their frequency, by itself it will not lead to such control. Persons diagnosed as carriers will have to know how to avoid parenthood. Improving the means of contraception and giving all parts of the population access to them are necessary second steps which will have to be taken if the primary knowledge of heredity is to be socially useful.

Knowledge of the mode of action of a deleterious gene, which leads to its detection in carriers, may also lead to knowledge of how to cure or alleviate the disease itself. Some phenylketonuric imbeciles appear to have benefited by having been put upon diets with small added amounts of the substance, phenylalanine, which, because of their genetic defect they are unable to utilize efficiently. With the progress of research, the conquest of some other hereditary diseases is certainly not beyond the bounds of possibility. If this sort of work has no other immediate effect, it will certainly change a prevalent medical attitude toward hereditary disease, namely, that there is little that can be done for those who are "born that way." Such an attitude neglects an important result of genetic research, the proof that in many cases hereditary diseases, like other hereditary characters, represent the norm of reaction of the genotype to the conditions in which it is placed. The problem of medical science is to find the conditions which minimize those reactions which are adverse to the health and well-being of the person.

A continuing problem for society in this field will be a two-fold one—two-fold because the biological character of each human being is the outcome of interaction between his heredity and his environment. As our knowledge of human heredity grows, so will our knowledge of what environments are suitable for possessors of particular genotypes. We have this knowledge now for diabetes, for example, an hereditary disease which is relatively common. I think we already recognize the "right" of a known diabetic to have insulin; it is already almost a social duty to have a source of it in his environment. A baby born with hemolytic disease due to the blood-type incompatibility of his parents must have a suitable blood donor or blood bank in his earliest postnatal environment to provide a substitute for his own ineffective blood for the first days of life.

Eventually the prevention or control of hereditary diseases will be as much a part of good public health practice as sanitary control now is. Parental instruction in the methods of contraception will be an accepted part of the service of the public health nurse or physician when hereditary disease is diagnosed in a child. The control of disease transmitted by infection has been acknowledged to be a public responsibility by most present-day societies; the societies of the future will have the same attitude toward those transmitted by genes.

But instruction and control of this sort, and provision of means for allaying the symptoms of hereditary disease when it appears, are only the most obvious of possible modifications of the environment. If increased ability to adapt by using our wits is the best hope of the future for human evolution, then the provision of varied opportunities for putting the great variety of human minds to use will be an important goal of future societies. Hardy, Weinberg, and their successors have shown us that human beings are never going to be uniform. Minds as well as bodies may be expected to retain their great variety. Any system of education based upon the assumption of uniformity will defeat its ends.

EFFECTS OF CULTURAL CHANGES

One of the results of the development of a sound theory of population genetics during the last few decades is the realization that the social and cultural forms which man has evolved have the power to affect his biology. We have seen a few examples of this in the discussion of isolating factors in Chapter VI. The question must have occurred to many readers at that time: what will happen if the rates of social, economic, and cultural change, obvious in the recent past, accelerate in the future? Granted that human variety will persist, what forms will it take? What will the future population of the world be like, biologically?

The facts which lead to such questions are, first, the rapid development of technology based on science, which began in Western Europe and America and spread to other parts of the world. This development is often referred to as the Industrial Revolution. The changes wrought by scientific advance have wider effects than on methods of production alone, but the first effects were on these methods. The most obvious biological effect of this has been an explosive increase in the world's population, dramatized by the title of a popular book, *Standing Room Only*. Much of the growth took the form of increase in size of urban centers and a shift, in the United States, for example, in the ratio of city dwellers to rural inhabitants. This goes by the name of urbanization. Along with its technology, other aspects of European and American culture spread to other regions; it is said that these regions are becoming "Westernized," so some of the effects which followed these changes in Europe and America may be expected to occur elsewhere.

Apart from the increase in numbers of people, which is now affecting such already dense populations as those of India, China, Egypt, and Indonesia, other changes will occur as the forces which mold human variety into patterns are modified. Some of these changes are already apparent. Geographical isolation has lost some of its power in keeping the gene pools of the continental races apart, although it still keeps the relative probability of intermarriage between, say, Asiatic and African lower than that for intraracial matings. As geographic barriers decline, however, new causes of isolation appear, and these modulate the tempo of hybridization. Transplanted peoples tend to retain linguistic and other cultural differences from the surrounding population and these reduce the rate of gene flow. The growth of cities tends toward amalgamation, it is true; but this is a slower process than some think. We still have our Harlems and our Chinatowns; and in cities such as Paris immigrants to the metropolis still cluster in parishes and neighborhoods with others of similar provincial or national or even village origins. When intermarriage does occur in the cities (and it is certainly increasing) it simply breaks old patterns of variety common to groups and emphasizes and increases individual variety. A rider on the New York subway will hardly believe that the metropolis diminishes human variety.

If we want some preview of what is likely to happen on the larger stage, we may turn, as scientists usually do, to smaller models in which some of the elements may be identified and evaluated. What, for example, does movement to the cities do to the gene pools which had been formed in smaller communities? In some individual European countries this has had the effect referred to as "isolate-breaking." It is marked by a decline in the frequency of cousin marriages, as larger circles of potential mates become available. There is a dramatic example of this in the European population of small rural Brazilian parishes which were subsequently incorporated into spreading urban areas like that of São

Paulo. The cousin-marriage rate, as documented by dispensation records in the parish marriage register fell from 20 percent or more in the early nineteenth century to around 5 percent in the recent records. In Sweden the fall has been less marked but appreciable in the less isolated areas. In many Japanese villages less affected by industrial developments, it has remained high, 20 to 30 times as high as the probable average in the United States. In India, residence in a large city such as Bombay does not by iteslf reduce the cousin-marriage rate since the Hindu peoples belong to mating communities separated not by geographic barriers but by caste rules. These apparently retain their force even when members of different castes live in close association with each other.

Much depends, evidently, on the social rules by which people live. If they all belong to one religious group, such as the state Lutheran Church of Sweden or the Catholic Church in Brazil, isolates based on distance may break down; if they adhere to different rules, as members of different caste communities, physical distances are of less importance.

Suppose the isolates do break up, as they appear to be doing in Sweden and some other European countries. Random mating then extends over a larger circle of probable mates. It is equivalent to an increase in the size of the intermarrying group, an expansion of the gene pool. Decline in the cousin-marriage rate is one sign of this; evening out of the local differences in the gene frequencies is another. In a population divided into small local marriage circles, a rare gene like that for juvenile amaurotic idiocy will have a spotty distribution. Some villages will have some heterozygotes in them, others none. Homozygotes with the disease will appear only in the former. When the heterozygotes are distributed more evenly, the chance that two of them will be marriage partners is reduced and the frequency of the disease will decline. Breaking of the isolates has, thus, a eugenic effect resulting not from deliberate acts of men but from changes in the social and economic structure. But it is only a slight effect, since reduction in the proportion of homozygotes has only a negligible effect on the frequency of the gene. As the gene becomes more evenly diffused, in the cities, for example, any subsequent change in the direction of increased inbreeding would again lead to increase in the incidence of the disease. If there are many different deleterious genes in the population, each kind individually rare, like those newly arising by mutation, then prohibition of the marriage of cousins over the whole population would effect some diminution in the incidence of hereditary disease.

The effect would not be large with mutation rates as they are at present. But, if the biological environments in which we live are going to be subjected to increased instability from radiation, some attention will have to be given to safe limits of inbreeding as well as to limits of radiation hazard. We know now that matter itself may be made unstable and that the by-products of atomic fission are powerful agencies for increasing the mutation rates of genes, human genes as well as those of the experimental plants and animals on which the effects of increased radiation have been measured. The danger is that there may be no lower limit to this effect, any increase in radiation increasing the probability of mutation. Since most new mutations are deleterious and recessive, the effects of these on the fitness of the population will be determined by the breeding structure of the population. Close inbreeding in isolates would give the maximum opportunity for homozygotes to appear. If, however, certain conditions should confer some advantage on the heterozygotes, these might be expected to increase, and a population with a breeding structure which gave the maximum opportunity for the formation of heterozygotes to arise might tolerate such genes. At present we do not know enough about the ratio between these two possibilities, one in which the gene is subject to selection by its effect on heterozygotes, the other by its effect when homozygous. It is not even certain that these are alternative possibilities since there is some evidence that selection in the long run acts on complexes of genes, on whole genotypes rather than on the effects of single alleles. Two general statements can be made: 1) the effects of changes in mutation rate will be determined in and by the population; 2) we do not yet have the body of fact or theory about this for human populations on which predictions can be based. In such a position, the counsel of wisdom would be not to permit increases in the risk known to accompany increased radiation until the extent of the risk can be determined.

The effects on human evolution of other technical and scientific changes lie largely in the domain of speculation. I have already indulged in some of this in the discussion of the future prospects for one gene, that for sickle-cell anemia. I shall conclude by quoting the last paragraph of a generally speculative lecture prepared in 1955 for a series with the provocative title "2000 A.D." My assignment was "The Prospects for Genetic Improvement,' which I interpreted to mean the prospects for continued forward evolution. "Wherever improvements, past or prospective, appear, they are associated with (or I suppose judged by) increased control and increased efficiency or effectiveness of scientific method, and this constitutes empirical knowledge, if not understanding. On the other hand, the other side of the ledger shows, on balance, the greater precariousness which is to be, for better or for worse, the condition which will dominate the existence of our descendants in 2000 A.D. That is a condition which ignorance, unreason, prejudice, even lack of rapid scientific advance, could quickly make entirely untenable for civilized man. Once we have got the jump on nature, we shall have to keep on jumping faster and farther. This is to say that we shall become increasingly more dependent on our wits, those specifically human attributes such as tool-using, and the use of language with its corollary, thought—in short on cultural development. Surely that is no novel conclusion. Twenty-five hundred years ago an anonymous preacher told his people: 'Wisdom is the principal thing; therefore get wisdom; and with all the getting, get understanding.' Today we should remember that more of that task lies ahead of us than behind us."

A Reference List

Auerbach, C., 1961. *The science of genetics*. New York, Harper and Brothers. A sound introduction to the elements of modern genetics, clearly written for the layman.

Beadle, George W., 1963. *Genetics and modern biology*. Philadelphia, American Philosophical Society. A short, clear account of the latest developments in genetics, including the deciphering of the code and its relation to human evolution.

Boyd, Wm. C., 1950. *Genetics and the races of man*, pp. 453. Boston, Little Brown.

Boyer, Samuel H., ed., 1963. *Papers on human genetics*. Englewood Cliffs, N. J., Prentice-Hall, Inc. Thirty-three of the original scientific papers published in this field since 1902—the sources of much of our information about human genes and methods of thinking about them.

Coon, C. S., S. M. Garn, and J. B. Birdsell, 1950. *Races*. Springfield, Ill. Charles C Thomas. A brief, speculative and stimulating discussion of possible effects of natural selection in the formation of human races.

Dobzhansky, T., 1962. *Mankind evolving*. New Haven, Yale University Press. The whole range of modern thinking about biological evolution focused on man and culture.

Dunn, L. C., 1965. *A short history of genetics: the development of some of the main lines of thought 1864-1939*. New York, McGraw-Hill Book Co.

Goldschmidt, E., ed., 1963. *The genetics of migrant and isolate populations*. Baltimore, Williams and Wilkins Co. Papers, mostly technical, by many specialists.

Ingram, V. M., 1963. *The hemoglobins in genetics and evolution*. New York, Columbia University Press. An up-to-date account by a biochemist.

Mather, Kenneth, 1964. *Human diversity.* Edinburgh and London, Oliver and Boyd. The nature and significance of differences among men.

Mourant, A. E., A. C. Kopec, and K. Domaniewska-Sobczak, 1958. *The ABO blood groups: comprehensive tables and maps of world distribution.* Oxford, Blackwell Scientific Publications.

Mourant, Arthur E., 1954. *The distribution of the human blood groups,* pp. 438. Oxford, Blackwell.

Penrose, L., 1960. *Outline of human genetics.* London, Heinemann. A brief nontechnical introduction to the basic facts, ideas, and methods of human genetics.

Race, R. R. and Ruth Sanger, 1962. *Blood groups in man,* 4th edition, pp. 456. Oxford, Blackwell Scientific Publications.

Schull, W. J., ed., 1963. *Genetic selection in man.* Ann Arbor, University of Michigan Press. Recorded discussions by 27 specialists on natural selection, operating on human populations through fertility, fecundity, differential mortality from disease, and similar topics.

Simpson, G. G., 1949. *The meaning of evolution.* New Haven, Yale University Press. The idea of evolution described with emphasis on its intellectual and philosophical implications.

Stern, Curt, 1960. *Principles of human genetics.* Second Edition. San Francisco, W. H. Freeman and Co. A thorough text book with extensive discussion of most of the questions dealt with in "Heredity and Evolution in Human Populations."

Stern, Curt, and Eva Sherwood, eds., 1966. *The origin of genetics: a Mendel source book.* San Francisco, W. H. Freeman Co. A new translation of Mendel's classic paper of 1866 with two of the papers reporting the rediscovery of Mendel's principles.

Index